1,000 THINGS TO LOVE ABOUT
America

ALSO BY BARBARA BOWERS, BRENT BOWERS, AGNES HOOPER GOTTLIEB, HENRY GOTTLIEB

1,000 Years, 1,000 People: Ranking the Men and Women Who Shaped the Millennium

Celebrating the Reasons
We're Proud
to Call the U.S.A. Home

HARPER

NEW YORK · LONDON · TORONTO · SYDNEY

1,000

THINGS

TO ★ Love About

AMERICA

BARBARA BOWERS, BRENT BOWERS,
AGNES HOOPER GOTTLIEB, HENRY GOTTLIEB

HarperCollins books may be purchased for educational, business, or sales promotional use. For information please write: Special Markets Department, HarperCollins Publishers, 10 East 53rd Street, New York, NY 10022.

FIRST HARPER PAPERBACK PUBLISHED 2010.

Designed by Janet M. Evans

Library of Congress Cataloging-in-Publication Data is available upon request.

ISBN 978-0-06-180628-5

10 11 12 13 14 OV/RRD 10 9 8 7 6 5 4 3 2 1

For
Our
Parents

INTRODUCTION

ENOUGH ALREADY. We Americans have been beating ourselves up for so long that we sometimes wonder if our luck is running out.

Think about the first decade of the 21st century: 9/11; wars in
Iraq and Afghanistan; Katrina; $4 a gallon gas; and, the worst reces-
sion since the Great Depression.

But we are a nation of optimists. We look on the bright side. And
despite the difficulties of recent years, there is plenty to celebrate
about America and its people.

Twelve years ago we wrote a book that stirred interest about the
best thinkers and doers of the second millennium: *1,000 Years, 1,000
People: Ranking the Men and Women Who Shaped the Millennium*. The
book did exactly what we wanted it to do: present the movers and
shakers of the years between 1001 and 2000 in a lively way that
made history accessible to a wide audience. More important, it
sparked lots of debate. Some readers called us brilliant. Others?
Well, we're still getting flak years later for failing to include Nikola
Tesla in the ranks of great physicists or to give credit to those re-
sponsible for the Internet.

We welcomed the feedback and congratulated ourselves on igniting a hot debate about people and history and calling attention
to the contributions of the famous and the obscure.

Now, what we did for the millennium, we want to do for America.

We spent 18 months tracking down and pinpointing the wonders that define who we are as a nation and as a people. We investigated institutions, principles, inventions, food, cultural landmarks,
parks, tourist destinations, companies, and ideas that represent the
best America offers.

In our earlier book we were guided by the "Biograph" system that measured a person's contribution—both positive and negative—to the millennium. For our current effort we affectionately coined the term "Amerigraph" to measure excellence among thousands of potential entries, narrowed the list down to 1,000, and undertook the perilous task of ranking them in their order of importance by asking:

- Does this rise to a level of national importance?

- Is it a best thing? (Sometimes we'd come across items
  that are incredibly important, but in a negative way).
  Is it something we love?

- Does this item have lasting influence and staying power
  or is it a passing fad?

- Does it contribute to the total of wisdom and beauty in
  the world?

- Finally, does it help define who we are as a people so
  well that people in other countries see it and say,
  "That's America"?

We included objects and concepts that are not uniquely American if we could point out that these items—like public school education or our newspapers—contributed to the sum total of what makes us so proud to be American.

Some of the ground rules: no people, only things. Bing Crosby is out, but his "White Christmas" recording is in. We focused on articles of everyday life and tried to craft a list that was current, not historical. The Battle of Gettysburg is out; the battlefield that 1.8 million people visit annually is in. We gave a nod to history by including the enduring principles of American democracy and some of the quintessentially American works of literature.

We eliminated most everyday objects like lightbulbs and pens that are ubiquitous, and concentrated instead on new ideas and items that are still changing the way we live in positive ways.

We want to engage America in a discussion about the best things the country has to offer in a way that will renew our pride. After all, for all the recent turmoil, we've gone through a lot worse and emerged stronger than ever.

1,000 THINGS TO LOVE ABOUT
America

1000. The Third Millennium: *The call of destiny.* It got off to a scary start with the terrorist attacks, two wars, venomous politics, and the Great Recession. But we've overcome worse in our 234 years, all the way from winning the Revolution to defeating totalitarian foreign powers to brushing off the now all-but-forgotten Y2K crisis that alarmists hyped as a potential economic Armageddon. Some pundits say China, India, or a united Europe will overtake America as the dominant economic player in the 21st century. Who can  tell? We do know that no nation will rule supreme in any sphere without sharing the ideals of freedom that make America great. Our hope is that by the year 3000, all the peoples of the world will have discovered how to live together in peace as one human family.

999. The Phrase "One of the Greatest Things About America": *All-purpose preface.* Do you love pinball machines? Well, then, one of the greatest things about America is pinball machines. Do you hate America? Not to worry; one of the greatest things about America is your right to trash it. As for us, we firmly believe that one of the greatest things about America is "one of the greatest things about America."

998. Monopoly: *Life's lessons on a board.* Stay out of jail, do community service, pay your bills, amass real estate, and collect your rents. Monopoly, with more than 200 million games sold in 37 languages, is the most popular board game in the world. It was created in 1934 by Charles B. Darrow of Germantown, Pennsylvania, who marketed homemade copies at a Philadelphia store, then sold the rights to Parker Brothers in 1935. We all have our favorite properties, named for locations in Atlantic City. Monopoly teaches children the ultimate lesson: the person with the most money at the end of the game wins.

997. The Indy 500: *Car racing's big one.* The Memorial Day weekend event in Indiana's capital, first held in 1911, is one of the first, probably the most famous, and almost certainly the biggest racing spectacle in the world, drawing an estimated 400,000 fans to watch 33 cars roar around the Indianapolis Speedway and cheer the one that covers 500 miles first.

996. The California Sea Lion: *Bark this way.* They roam waters from Canada to Mexico, but the noisiest population frequents at Pier 39 in San Francisco. There, hundreds of these massive hulks snooze in the sun or jockey for space on crowded harbor rafts, protected from hungry sharks. Sea lions have doglike faces, are intelligent, social, playful, and increasing in numbers. They're the ones usually tapped to entertain at zoos and aquariums nationwide.

995. Sweet Potatoes: *AKA yams.* There are botanical differences between sweet potatoes and yams, but Americans use the terms interchangeably. The sweet potato is a New World plant encountered by explorer Christopher Columbus perhaps as early as 1492. Slaves in the American South called the sweet potato "nyamis"—the name evolved to "yams"—because it looked like a vegetable from their homelands. This superfood, packed with vitamins and minerals, enlivens pies, puddings, muffins, candied vegetables, and biscuits. The best serving? Microwaved, whole.

994. The Music of Hank Williams: *We can't help it if we're still in love with him.* He came out of Alabama with a guitar, a twang that seemed filtered through the dust of southern roads, and a gift for songwriting. Before his death at 29 in 1953, burned out by drink, drugs, and success, he left some of country music's unrivaled hits: "Lovesick Blues," "Hey Good Lookin'," "Your Cheatin' Heart," "I Can't Help It (If I'm Still in Love With You)," "Cold, Cold, Heart," and "Jambalaya," recorded by other artists more than 500 times.

993. Napping at the Office: *Mental health moments.* zzzzzzzz zzzzzzzzzzzzzzzzzzzzzzzzzzzzzzzzzzzzzzzzzzzzz Zzzzzzzzzzzzzzzzzzzz zzzzzzzzzzzzzzzzzzzzzzzzzzzzzzzzzzzzzzzzzzzzzzzzzzzzzzzzzzzzzz zzzzzzzzzzzzzzzzzzzzzzzzzzzzzzzzzzzzzzzzzzzzzzzzzzzzzzzzzzzzzz zzzzzzzzzzzzzzzzzzzzzzzzzz. Wha? Huh? Oh? No, I was just thinking with my eyes closed.

992. The Tuxedo: *Shortcut to high fashion.* As a lark, tobacco magnate Pierre Lorillard IV's son Griswold and his buddies cut off the tails of their formal black coats to wear at a ball in the exclusive Tuxedo Park enclave north of New York City in 1886. The style caught on, and now is an international symbol of sartorial elegance. For ordinary folks, the crux of the tux question is: buy or rent? Experts disagree, but "renting offers more options to men whose size may change," opined garment guru Francine Parnes.

991. Duct Tape: *Ties that bind.* First applied to heating and air-conditioning ducts, this vinyl, fabric-reinforced, pressure-sensitive tape is a staple in American households. Developed during World War II, it was used to seal ammunition cases and repair jeeps, weapons, and planes. Folks soon found plenty of other uses (including, alas, covering crime victims' mouths). Superstrong and cheap, this material actually draws a cult following: people join duct tape clubs, and even make wallets and prom dresses with this stuff.

990. Window Screens: *First line of defense against insects.*
This ultrasimple nonpolluting device, made of wire mesh and a metal or wooden frame, keeps bugs where they belong: outdoors. That might seem obvious, but most of the rest of the world has yet to discover the low-tech breakthrough.

989. Our Gosh-Gee-Whiz Openness: *Why the world hates us and loves us at the same time.* Many is the European sophisticate who has scoffed at our country-bumpkin ways, only to acknowledge our greatness upon reflection. Some even admire our idealism and readiness to use our wealth, scientific brainpower, and military might to defend the Western way of life. "We are inclined, in our snobbish way, to dismiss the Americans as a new and vulgar people, whose civilization has hardly risen above the level of cowboys and Indians," British journalist Andrew Gimson wrote recently. "Yet the United States of America is actually the oldest republic in the world, with a constitution that is one of the noblest works of man." Why, thanks, fella! And have a nice day!

988. Alternative Medicine: *Pills to pop when you're tired of popping pills.* We ingest flax seed oil for overall good health, echinacea to ward off colds, glucosamine to lubricate our joints, ginseng to improve our memory, fish oil to lower cholesterol, rose hip to look younger, and Saint-John's-wort to lift our spirits. Do they work? Who knows? But a recent federal survey showed that more than a third of adult Americans turn to dietary supplements, herbal remedies, and alternative therapies like yoga and acupuncture in lieu of or in addition to regular visits to the doctors.

987. Martinis: *The best way to welcome each new year.* We recommend a dry, stirred gin martini with at least three olives. Or, for the weak of heart, an apple martini with a maraschino cherry garnish to provide a festive toast on January 1. One among us suggests hors d'oeuvres of anchovies on toast to help sop up the alcohol.

986. Legendary Creatures of the Wild: *Balm for boring summer nights.* Sure, almost every country has a few (like Scotland's Loch Ness monster), but the American versions are notably picturesque. Bigfoot, also known as Sasquatch, the big, hairy humanoid that roams the Pacific Northwest, is the most famous. The Jersey Devil, the flying biped that haunts the Pine Barrens of southern New Jersey, is the scariest. And Mothman, a web-footed hunk (is he gay?) with an insect head, giant moth wings, and a penchant for badgering motorists in rural West Virginia, is the most ridiculous.

985. Comedy Clubs: *Funny stuff.* That hilarious entry about legendary creatures of the wild is a tough act to follow, so we're going to have to struggle to make this work. We guess the first comedy club in America was a hut in the Jamestown colony and the jokes were about starvation and savages. "I love Indians. Take Powhatan. Please." Comedy clubs were nightclubs until they got rid of the expensive musicians and kept the comics and the drunks. If you are in a strange town and looking for entertainment, Google "best comedy club." In Milwaukee you will learn that Tommy Chong, Drew Carey, Rita Rudner, and Elayne Boosler played at JD's Comedy Cafe on East Brady Street. You might see a rising star just as good. Or the headliner could be a dentist as funny as a root canal or a hack dropping "F" bombs to get a pulse from the last two people in America who think dirty words are funny if you say them loud. Thank you. You've been a great audience. Give ChapStick a warm welcome and God bless.

984. ChapStick: *Balms away.* It would be a wonderful world if every ailment could be wiped out as easily as chapped lips. The Virginia pharmacist who invented ChapStick in the 1870s didn't have some of the chemicals that are in today's version, but a few old standbys remain on the list of ingredients such as camphor, wax, and mineral oil. ChapSticks work and they make great Christmas stocking stuffers.

983. American TV Cartoons: *What's up, Doc?* We gave the world Bugs Bunny, Bullwinkle, Tom and Jerry, Magilla Gorilla, Scooby Doo, Fred Flintstone, Barney Rubble, George Jetson, Tommy Pickle, Doug Funny, Squidward, Homer Simpson, and Huey, Dewey and Louie. Back in the day, the best time to be a kid was Saturday mornings, which were packed with back-to-back cartoons. Now, you don't have to wait for Saturday: the Cartoon Network supplies a seemingly endless broadcast of our animated favorites for the kid in all of us.

982. The Old Farmer's Almanac: *Read it and reap.* North America's oldest continuously published periodical dates from 1792. Once required reading in outhouses, the familiar yellow-covered booklet provides traditional tips on gardening, folkloric features like "Jeepers, Creepers, It's the Peepers," and long-term weather forecasts claiming 80 percent accuracy. Its tide charts are so spot on that Washington almost banned them during World War II for fear they'd help German spies.

981. The Great Alligator-Python Battle: *Gory glimpse of Florida's changing wildlife.* All sorts of exotic species have invaded the state, including the giant Nile monitor lizard. Recently, pythons—which can grow to 20 feet and 200 pounds—have spread through the Everglades and beyond, their numbers now in the hundreds of thousands compared with 1.5 million alligators. In January 2003 a crowd gathered to behold an epic, 24-hour fight between a python and an alligator trapped in a mutual death grip of coiled muscle and razor teeth. The alligator won.

980. Minuteman Missile National Historical Site: *Paging Dr. Strangelove.* Feeling nostalgia for the simpler days when Mutu-

ally Assured Destruction was U.S. strategy against Soviet might? Visit an underground ICBM launch control center and missile silo near Wall, South Dakota. "Learn how nuclear war came to haunt the world," the brochure says.

979. Leftovers: *A refrigerator's hidden treasures.* Just when you thought there was nothing left in the house to eat, your fridge revealed a four-day-old hamburger, two spoonfuls of pudding, a small helping of linguine with clam sauce, one slice of Mom's meat loaf, and enough mashed potatoes to feed a Little League team. We Americans make huge  quantities of food and throw nothing out because of guilt about all the people who go to bed hungry. Our solution is to eat this old stuff or keep it in the fridge until it turns green, when we can then toss it without even slightly tweaking our conscience.

978. OK: *World's most repeated word.* It's also spelled O.K., okay, and even okey, OK? Inhabitants of the tiniest village in the most remote rain forest utter it. You can't escape it, but why try? It has a Yankee Doodle Dandy ring to it, though nobody has been able to trace it definitively to its source.

977. Doggie Bags: *Remembrance of meals past.* People used to tell waiters to wrap up leftovers for the dog. Now there's no more pretense. Half the restaurant portion you were too stuffed to eat tastes great reheated at home. You have to be selective. Leave the canned peas behind.

976. *The Catcher in the Rye*: *Best American novel never made into a movie.* This coming-of-age tale about troubled teenager Holden

Caulfield has made an indelible imprint on adolescent culture, with sales of 35 million copies since its publication in 1951. But its famously reclusive author J. D. Salinger fought ferociously to keep writers and film producers away from his chef d'oeuvre. It will be made into a film someday, though perhaps not in your lifetime.

975. Yogiisms: *The catcher in the wry*. Thoreau, Dewey, James, Chomsky? Mere backups to the wisdom of Yankees-baseball-great-turned-Yankee-philosopher, Yogi Berra. "When you come to a fork in the road, take it." "Nobody goes there anymore, it's too crowded." "You can observe a lot by watching." These are all *nonpareil* insights into *la condition humaine,* to be sure. Yet, somehow, it seems we have *déjà* heard them somewhere else.

974. Hawaiian Shaved Ice: *Cool down a hot day*. Sno-cones are for kids. This delightfully refreshing tropical treat satisfies a grown-up palette with flavors like Piña Colada and Mango Pineapple. Try the outdoor shaved ice stand at the foot of Diamond Head in Honolulu. Or, if Hawaii seems a bit far to travel for this cool refreshment, we also like the imitation ices on the boardwalk in Bethany Beach, Delaware.

973. The Outer Banks: *The Wright stuff*. These narrow barrier islands off the coast of North Carolina boast lots of other attractions, like Jockey's Ridge, the tallest natural sand dune on the East Coast, and the ruins of the lost Roanoke Colony, birthplace of the first American of English descent, Virginia Dare. But you *must* go there at least once to gawk at Kitty Hawk, birthplace of flight, where the Wright Brothers piloted the first powered airplane in 1903.

972. It Is What It Is: *The Serenity Prayer in brief*. The verbal equivalent of a shrug, the phrase "It is what it is" captures the American spirit of accepting what we cannot change. Who knows who said it first? It sounds like a Yogi Berra quip, but even *New York*

Times word sleuth William Safire couldn't find the origin of the phrase. Gary Mihoces of *USA Today*, who heard it ad nauseum from football coaches reluctant to dissect their teams' losses, called it "the all-purpose alternative to the long-winded explanation." New Yawkahs make do with "Whaddayagonnado?" social snobs with *"C'est la vie,"* but we prefer "It is what it is." Don't ask us why. It is what it is.

971. No-Questions-Asked Return Policies: *A second chance for people with second thoughts.* The practice is, of course, abused by cheaters who pull tricks like buying designer clothes on Friday, wearing them on Saturday, and returning them on Monday. That said, it is a relief for the rest of us to get our money back without fanfare for merchandise that turned out to be defective—or just not quite what we wanted. No other country comes close to matching this largesse.

970. The London Bridge: *Most beautiful British bridge in America.* Built in 1831 over the Thames but unable to support the weight of 20th century traffic, it was bought by oil magnate Robert McCulloch in 1968 and moved to his beloved Lake Havasu City, Arizona, as a tourist attraction. It was the setting for a 1985 made-for-TV horror film called *Bridge Across Time*.

969. Internet Shopping: *Yet another distraction at work.* You can look busy at the office while you are actually buying your weekly food order, completing your Christmas list, or comparison shopping for a new refrigerator. The United States leads the way in Internet shopping—with sales hovering around $150 billion.

968. Halloween: *A frightfully over-the-top holiday.* Other countries ignore the eve of All Saints Day, but in the United States, October 31 is every child's dream. The amount of candy you amass

trick-or-treating door-to-door is only limited by your stamina. Halloween was almost wiped out in the 1970s during a tainted candy scare, but it surged back even stronger. In fact, what used to be a children-only event has morphed into an adult-costume celebration. Check out the anything-goes Greenwich Village Halloween Parade in New York City for some really kinky dress-up.

967. Cheap Gas: *Shut up and drive.*

When gas hit $4 a gallon in the United States, it averaged around $10 in Western European countries. Another reason to stop whining about America. And to travel by train when you're abroad.

966. Toilet Paper: *Designed to do a dirty job.*

We used to travel abroad with a roll of Scott's toilet paper in our suitcase. The situation has improved, but we still count America's ability to make and sell a decent roll of toilet paper as the envy of the thrones of Europe. You can only appreciate how good we have it when you've seen how rough it can be elsewhere.

965. Chocolate Bunnies: *Easter treat that says a lot about your psyche.*

If you eat the ears first, you are sexually repressed. If you bite off the tail and then methodically make your way through the torso, you have a fear of commitment. If you cut it into pieces and serve it in a candy dish, you have a deeply rooted inferiority complex. If you let it melt on the backseat of your car, you are just stupid.

964. E-ZPass: *Ticket to ride.*

A shout-out of thanks to millions of paranoid motorists who fear Big Brother and refuse to use the best convenience on the road. We wave at drivers in long toll lines as we flash our E-ZPass and fly by. While there are other electronic toll collectors (iZoom in Indiana, SunPass in Florida, or FasTrak in California, for example) E-ZPass is the most widely accepted one. It collects tolls along the Northeast corridor on our most heavily traveled bridges, tunnels, and roads.

963. Lawn Ornaments: *Dazzling displays of kitsch.* Pink flamingos, ceramic frogs, strutting geese adorned in human clothing, elks with reflector globes in their antlers, cement statues of dogs—the outdoor adornments that have sprouted up in suburbia as nowhere else in the world are so vulgar they might actually be hip. Our favorites glimmered only feet from the front stoop of the home where two of us lived for years: three-foot-tall hollow plastic gnomes that lit up at night.

962. Isabella Stewart Gardner Museum: *A real steal.* Strolling through the galleries that overlook the sun-drenched, flower-bedecked courtyard of this Venetian-style palazzo in Boston is as delightful as gazing upon the artworks. What a collection: Renaissance, 19th-century American and Impressionist paintings, and sculptures, tapestries, furniture, decorative art, and manuscripts. Sadly, it was the scene in 1990 of the biggest art theft in U.S. history, treasures worth $500 million, including paintings by Vermeer, Rembrandt, Degas, and Manet. Blank spaces mark the spots where they hung. Founded in 1903 by an heiress to a linen fortune, the museum admits women with the first name Isabella for free.

961. The Name "America": *An over-the-topography debate.* Most historians trace it to 15th century Florentine explorer Amerigo Vespucci, but sundry scholars and cranks have proposed alternatives. They include the gold-rich Amerrique region of Nicaragua; the Old Norse word Ommerike meaning "farthest outland," a Welsh merchant named Richard Amerike who helped finance John Cabot's 1497 voyage to the New World, and a Mayan word that sounds like "America" and means "land of the wind." We'll stick with Amerigo, but whatever the origin, the word fires the imagination like no other.

960. Trucking Culture: *The long haul.*

Nearly 2 million drivers make a living hoisting themselves behind the wheel of their 18-wheelers and heading out for the great yonder. Fiercely independent, they ride the asphalt range subsisting on talk radio, coffee, and—so the myth goes—an intense longing for their sweethearts back home. Their odysseys are the stuff of country hits like "Six Days on the Road" ("Well, I pulled out of Pittsburgh, Rollin' down the Eastern Seaboard . . ."), "Convoy" ("We's headed for bear on eye-one-oh, About a mile outta Shakeytown"), and "Rolaids Doan's Pills & Preparation H."

959. Literary Feuds: *Highbrow spectator sport.*

Just as gossip columns provide grist for ordinary people's schadenfreude in the mishaps of Hollywood celebrities, so public spats between literary lights let the intelligentsia exult in the petty bickering of cultural giants. We think of Mark Twain describing Bret Harte as "an invertebrate without a country"; Gore Vidal calling William F. Buckley, Jr., a "pro-crypto-Nazi" in a TV debate, and Buckley shooting back, "Now listen, you queer"; and our favorite, Mary McCarthy's putdown of Lillian Hellman: "Every word she writes is a lie, including 'and' and 'the.'"

958. Spam: *Food for hard times.*

Hormel Meats first produced this compressed pork and minced ham combo in 1937 and it became famous in World War II when 15 million cans a week were shipped to Allied soldiers. "Without Spam, we wouldn't have been able to feed our army," Soviet premier Nikita Khrushchev recalled. Spam gets a bad rap from nutritionists; eight ounces contain all the salt and fat an adult should eat daily. But Spam remains a popular food for the budget conscious. When the recession of 2008 hit, production went into overtime and Hormel profits soared.

957. State Quarters: *A reason to save $12.50.*

We had a grand time squirreling away $1.25 a year between 1999 and 2008. We've now got a full set of 50 pristine quarters with designs on the back from each of the 50 states. In moments of financial desperation, that's comforting.

956. Dewey Wins: *The image that keeps opinion surveys honest.* The sight of a grinning president-elect Harry S Truman holding up the *Chicago Tribune*'s DEWEY WINS banner headline about the 1948 election, perhaps the most famous American newspaper photo, appears in the nightmares of political pollsters and makes them recheck their numbers over and over. Alas for them, their back-and-forth declarations of who won Florida—and thus the presidency—in the 2000 presidential elections made their nightmares come true.

955. Healthy Disrespect for Elected Officials: *Geez, I could do as good a job as that boob!* Since the beginning of the Republic, Americans have believed that any—well, almost any—citizen could do as good a job in office as the person who was elected. That belief helps keeps the pressure on incumbents who view themselves as indispensable. And it explains the annual pile of write-in ballots for Mickey Mouse.

954. King-Size Beds: *Prairie-size home companions.* We know a couple who bought a full-size bed when they married, switched to a queen five years and two kids later, and celebrated their  30th anniversary by buying a 76-by-80-inch king to end nocturnal shoving. They have grown closer by moving apart.

953. Nobel Laureate Only Parking: *Another benefit to being really smart.* The University of California, Berkeley, maintains a bank of parking spaces reserved for Nobel Laureates. On a campus

where parking is at a premium, it could be one of the best perks for winning the coveted prize. Sixty-one UC Berkeley faculty and alums have won a Nobel prize. There's a $55 fine if you are so presumptuous as to park in one of the spots without the testimonial to back you up.

952. Our Foot Fetish: *Just saying no to the metric system.* Some of us
remember when half-pint bureaucrats tried to wean us of our love affair with imperial measurements like miles and quarts and pounds, but we would have
none of it. Sure, liters and grams have made limited headway on supermarket shelves, but quick: how tall are you in centimeters? It takes no more than an ounce of sense to know the old ways are best. Critics scoff that only Liberia and Myanmar stand with us in defiance of metric units, but what kind of yardstick is that?

951. Small Town Orchestras: *Giving Bach to the community.*
"Without music, life would be a mistake," the gloomy metaphysician Frederick Nietzsche declared. America agrees. Amateur and semiprofessional orchestras abound in the United States, playing Ludwig van Beethoven, Antonio Vivaldi, Charles Ives, and their ilk to chase away the blues and create oases of culture. Like at Colby Community College in West Kansas, where the Pride of the Prairie Orchestra rehearses on Monday nights and attracts audiences for three concerts a year.

950. USA Today: *Welcome mat in America's hotels.* Traveling
executives take comfort in opening the doors of their hotel rooms in nameless cities and greeting the morning with a copy of one of the top-selling newspapers in America. *USA Today*, dubbed "McPaper" by pundits who sneered it was the journalistic equivalent of fast food, actually revolutionized stodgy newspapers when it debuted in 1982. Its novel use of color and computer graphics, and a strategy to

partner with big hotel chains, helped USA Today grow as a national newspaper even when other papers were shrinking.

949. **Giving the Old College Try**: *Never say die!* Americans love a challenge, so we'll make a stab at solving difficult problems even if our attempts are likely to fail miserably. We call that "Giving it the old college try." Most likely an old baseball term (although its genesis is muddy), the "old college try" is the way we warn others that the outcome might fall short of our reach.

948. **Airstream Trailers**: *Down-market dignity.* Elitists speak of trailer trash, but even they are stopped short by the aesthetic perfection of these minimalist mobile homes. In a market marked by cheapness that can border on vulgarity, Airstreams' rounded aluminum architecture has a retro Depression-era aristocratic allure. Its popularity stems from so many all-American impulses—owning a home, hitting the road, being practical—that you could almost say it is an iconic icon.

947. **The Broadmoor Hotel**: *On a higher plane.* Known as the "Grand Dame of the Rockies," this five-star luxury hotel and resort near Pike's Peak in Colorado Springs sits more than 6,000 feet above sea level. The site offers a top-rated tennis facility and three 18-hole golf courses, including one used for the U.S. Open. In fact, the 2011 U.S. Women's Open will be played there drawing the greatest female golfers in the world.

946. **Birding**: *Nature's tweet.* Bird watching combines a couple of things Americans love: getting outdoors and competing. Spotting the 800 or so species found in the United States and adding the names to a life list is the birding equivalent of climbing Mount Everest—an odyssey that could start by encountering a pigeon in Times Square and end with a sighting of a Kittlitz's Murrelet off the Aleutian Islands. All it takes to start the hobby is a guide to local varieties and a good pair of binoculars. And don't feel intimidated on your first foray. Section 4(a) of the American Birding Association's Principles of Birding Ethics says, "Be especially helpful to beginning birders."

945. Allen-Edmonds Shoes: *Aaahhhh.* While American shoe manufacturers have been ruined by foreign imports, there's still one U.S. firm happily supplying fine stores in this country and overseas with quality footwear. Founded in 1922, this privately held company with 600 employees in Wisconsin and Maine produces pricey men's leather shoes (classic penny loafers run $285 a pair) that many extol as the finest anywhere for style and comfort.

944. Endless Ice Cream Flavors: *Cool choices.* Baskin-Robbins, for one, started over 60 years ago with 31 flavors. That seemed a lot then. Now they're offering over 1,000, with everything from Daiquiri Ice to Mango Tango. Against this explosion, even Cookie Dough looks classic. While it's always fun to experiment, we trend toward the tried-and-true in this $23 billion ice cream and frozen desserts market. Analysts say vanilla scoops up 30 percent of the U.S. ice cream market; chocolate, 10 percent; butter pecan, 4 percent; strawberry, 3.7 percent; and chocolate chip mint, 3.2 percent.

943. The Vermont Country Store: *Memories are made of this.* Yearn for Charms candy or Beeman's Gum? Nostalgic for wooden pickup sticks? Long for the scent of Lifebuoy, the original deodorant soap? Run by third and fourth generation shopkeepers, this quaint general store in Weston, Vermont, takes pride in tracking down hard-to-find products. Now whatever happened to Candy Dots?

942. Arabia Steamboat Museum: *Keeping the early West alive.* Opened in 1991, this Kansas City, Missouri, attraction displays

thousands of artifacts of the steamboat *Arabia*, which sank in the deep mud of the Missouri River in 1856, its only casualty a mule. *Arabia* held 222 tons of clothes, tools, food, and medicine bound for the frontier. The muck preserved everything so well that once the vessel was unearthed in 1988, excavators found jars of still-sealed fruits and vegetables, including a 132-year-old pickle that tasted just fine.

941. Mohonk Mountain House: *Enchanted hideaway.* Just 90 minutes north of Manhattan, this seven-story Victorian castle is one of our oldest family-owned resorts. Built in 1869 on the Shawangunk Ridge in the heart of the Hudson Valley, it boasts 265 rooms overlooking thousands of acres of unspoiled forest. Guests have included industrialists John D. Rockefeller and Andrew Carnegie, as well as presidents Teddy Roosevelt and William Howard Taft. History was made here, too. Between 1895 and 1916, negotiators met at Mohonk to create the Permanent Court of Arbitration in The Hague, Netherlands.

940. Colorado Fourteeners Initiative: *Saving the mountains.* Volunteers backed by public and private funds restore trails and undo environmental damage the half million visitors inflict on the 54 Colorado mountains 14,000 feet or higher. The initiative is one of hundreds of trail councils and conservancies that aim to keep America's open space from being trashed.

939. Hatch Show Prints: *Letterpress legends.* A Nashville, Tennessee, fixture since 1879, this old-style, crammed-to-the-rafters print shop still turns out distinctive posters promoting everything from a Shania Twain concert to Nike sneakers and Jack Daniel's. Years past, Hatch touted the talents of African-American entertainers like Cab Calloway, Bessie Smith, Duke Ellington, and Louis Armstrong. Even now, the eye-catching look scores with simple designs and bold lettering, fashioned from 10,000 basswood and maple-wood blocks, countless photo plates, and drawers of wood and metal type.

938. Ozark Folk Center State Park: *Living museum.* Located
in the beautiful Ozark region covering parts of Arkansas, Illinois,
Kansas, Missouri, and Oklahoma, this Mountain View, Arkansas,
spread is the only state park dedicated to preservation of southern
mountain folkways. The center offers folk music aplenty from ban-
jos, fiddles, dulcimers, and autoharps. There's jig and square dancing plus demonstrations of pioneer skills and crafts like
woodcarving, blacksmithing, soap-making, and broom-making.

937. Simon & Garfunkel Music: *Much better than the sound
of silence.* It's the kind of music to load on your iPod and let whisper
in your ear when you are fighting insomnia. In fact, a British study
showed cows that listened to the soothing classic "Bridge Over
Troubled Water" produced more milk! Musical maestro/lyrical poet
Paul Simon met his future singing partner, Art Garfunkel, in the
sixth grade at P.S. 164 in Queens, New York, in the 1950s. They were
a folk phenomenon of the 1960s and have been singing together off
and on ever since. Simon has had a wildly successful solo career,
while Garfunkel has dabbled in acting (remember *Carnal Knowl-
edge?*) and solo performances through the years. They get together
and tour every couple of years to remind us just how wonderful
their music is. Our favorites: "For Emily, Whenever I May Find Her,"
and "Homeward Bound."

936. Hair Salons: *Confessionals with
style.* Revealing sins to a priest is passé
and psychiatry is expensive. For the cost
of a haircut—and we recommend a hefty
tip—Americans can blab in hushed tones
about their infidelities, crimes, and medical news with confidence that their se-
crets will not be retold. We asked a stylist whether clients tell her
about their extramarital affairs. "Three times a week," she answered. "What I hear from the chair, stays in the chair."

935. Salad Bars: *Dieters' delight.* There's no need for people watching their pounds to stay home when the rest of the gang wants to pig out. There's usually something tasty at the salad bar, though taking too many bacon bits would be cheating.

934. Cedar Point Amusement Park: *Fast forwarding for humans.* It bills itself as the roller coaster capital of the world and says it has been voted best amusement park on the planet for 11 years straight. With rides like Corkscrew, Disaster Transport, and Wicked Twister, the park gets no argument from us. We get dizzy just looking at the pictures on its Web site.

933. University of Colorado at Boulder: *Conservation smarts.* Rated the top Green U.S. College by *Sierra* magazine, CU created the first student-directed recycling center, pioneered a program offering its 30,000 students free city bus passes, provides a free bike-share program, and uses biofuels in campus vehicles.

932. Acoma Pueblo: *Ancient dwelling.* A few Native Americans still live in this historic pueblo built atop a massive sandstone mesa 55 miles west of Albuquerque, New Mexico. Also known as "Sky City," it dates from the early ninth century and is considered the oldest continuously inhabited village in the United States.

931. Raspberries: *A resounding Bronx cheer!* New York is a tough crowd. And if you don't measure up to its high standards, the audience will let you know. We use the term "Bronx cheer" for the noisy splat that comes when you stick out your tongue and blow. The name comes from fans of the New York Yankees, a.k.a. the Bronx Bombers, who wanted to let players know they were unhappy. When babies make this distinct sound, we call it "raspberries."

930. Hell's Canyon National Recreation Area: *Despite the name, 652,488 acres of God's country.* Hiking the full 900 miles of trails in this spectacular rocky wilderness straddling the Oregon-Idaho border would be like walking from Chicago to Charleston, South Carolina. The canyon has the deepest gorge in North America, and the Snake River runs through it.

929. *The Clearwater*: *Environmental flagship.* Since folk-singing great Pete Seeger launched the Clearwater program on the Hudson River in 1969, more than half a million youngsters and hundreds of thousands of adults have sailed her for their first look at an estuary's ecosystem. The 106-foot-long sloop was among the first U.S. boats to conduct science-based environmental education onboard, a method now copied around the world.

928. Residential Styles: *Home sweet homes.* Americans relish variety, and we've got more home styles than you can shake a shingle at. Our estimated 130 million dwellings range from European-influenced classics like Greek Revival and Italianate to homegrown rustic Log Cabins, sprawling single-story ranches, pitched-roof salt-boxes, stately Victorians, stucco Spanish colonials, and artsy Craftsman bungalows.

927. Ice Cream Trucks: *You can always hope today's the day!* Will Dad finally say, "Yes!"? We have fond memories of the jingles that signaled the arrival of the ice cream truck on our streets. The distinct white Good Humor trucks of the 1950s and 1960s gave way to the blue Mr. Softee vehicle with the original cone-head mascot. Then came Little Jimmy Italian ice and today's more generic vans that distribute all kinds of prepackaged ice cream bars. Whatever the form, kids of every age light up when they hear the nostalgic tune that signals the arrival of the ice cream truck.

926. Singing Onstage with Barry Manilow: *We just can't smile without him.* This pop music idol and darling of middle-age grandmothers has been inviting women on stage to croon with him in concert for more than 30 years. Leave the "Pick Me, Barry!" signs at home. God bless 'im—he likes to call up the woman who was chosen last back when we were divvying up sides in gym class.

925. Belly Dancing Classes: *Dreaming of Jeannie.* For women who are tired of exercise classes run by skinny 20-somethings with no curves, learning to shake, rattle, and roll a woman-size torso is a trendy alternative. They learn to release their inner goddesses. And how about those veils?

924. Angus Beef: *It's what's for dinner, and lunch, and breakfast.* More than 27 million U.S. steers and heifers make the ultimate sacrifice every year, and most of them are Angus, the descendants of cattle imported from Scotland in the 1870s. Ranchers argue over whether the up-and-coming red variety is as hardy and tasty as the more venerable black Angus. By the time these animals turn into prime steaks, who cares?

923. Duck Decoys: *Fooling fowl.* Our nation's more than 1 million duck hunters endure mud, cold water, and wind to bag quarry attracted to their gun sights by hand-carved wooden decoys painted to resemble every type of bird from mallard to ringneck. These American carvings can reach the level of art. But we still prefer our decoys high and dry on a fireplace mantel—rustic touches for anyone who loves the outdoors.

922. Puerto Mosquito: *Floating in stardust.* This bay off Puerto Rico's Vieques Island is an enchanting natural wonder. There are up to 720,000 single-celled bioluminescent dinoflagellates—half-plant, half-animal organisms—per gallon of water. When agitated, the critters transmit greenish light. Take the plunge on a moonless night and you literally glow in the dark.

921. Almonds: *All they are cracked up to be.* California produces about 1.3 billion pounds of almonds annually, which adds up to about 75 percent of the world's crop. We fret that the mystery of the dwindling honeybees and the resulting lack of crop pollination could affect our favorite nut. We like almond paste (marzipan) in our morning Danish and string beans sautéed with almonds for dinner.

920. The U.S. Open: *Meaningful tennis in September.* It's a great diversion when your football team is at training camp and your baseball team is dangling at the bottom of the standings. Just when you can't think of a good reason to open the sports pages of the newspaper, along comes the U.S. Open. This annual tournament takes place at the National Tennis Center in Queens, New York, in late August, early September. It's the United States Grand Slam event that showcases the world's best tennis players.

919. *The Wizard of Oz*: *Better than Kansas.* Today's youngsters rarely enjoy a movie made in the 1930s, but this one they do. Dorothy, Toto, the witches, the cowardly lion, the men of tin and straw and those munchkins remain riveting 70 years after they cavorted on the screen. "Over the Rainbow" may be the best song ever performed in a film. Like a lot of classics, *Oz* was slow to recoup its costs. Not until a second release 10 years after its 1939 debut did the film earn the $2.77 million MGM spent to make it.

918. Prize Pigs: *Taking home the bacon.*

The largest pig on record was 1,600-pound Big Norm of Hubbardsville, New York, who died of an apparent heart attack in 2005. That's what happens when you eat like a pig. For porkers that win prizes at county fairs as their beaming handlers kneel nearby, size doesn't matter, though. It's all about appearance and the promise of a tasty breakfast.

917. Lowry Park Zoo: *Encounters of the closest kind.*

At this 56-acre Tampa, Florida, zoo rated tops by *Parents* magazine, kids can brush a goat's fur, ride a llama or camel, and feed snacks to rhinos, stingrays, lorikeets, and giraffes. The place couldn't be more child-friendly: it's the only U.S. zoo with an accredited preschool and kindergarten.

916. Dick's Drive-Ins: *Fast-food upmanship.*

Established in 1954, one year before the start of the McDonald's franchise, Dick's now has five Seattle-area outlets that thrive on the real thing. No prefab or machine-handled food here. Staffers manually slice blocks of cheese, hand dip and individually whip milk shakes, cut french fries from fresh potatoes, and broil the finest beef for burgers. Vive la difference!

915. American Familiarities and Nicknames: *BFF, but what's your last name again?*

If your name is Elizabeth, we'll call you Liz. If your name is Michael, it's "Mike" to us. We don't stand on ceremony. Waitresses call you "honey" and "sweetie." Department store clerks call you "dearie." On any given day you can be Bro, Sistah, Mack, Charlie, Gangsta, Sweetheart, Babe, Baby, Girlie, Babydoll, Mister, or Lady. People of other cultures are standoffish, while we like to get to know you, superficially, but well.

914. Sousa Marches: *Oom-pah-pah.* John Philip Sousa (1854-1932) left an oeuvre of rousing band music and marches that continue to inspire us on the football field, at patriotic events, and at middle school band concerts. The melody to his "Stars and Stripes Forever" provides the music for the childhood favorite "Be Kind to Your Web-Footed Friends." He has celebrated the Army, the Navy, the Marines, the *Washington Post*, the state of Minnesota, the Wolverines, and countless others in song.

913. Owning a Horse: *Whoa.* Americans don't have to live in the sticks to own one of the country's 5 million horses. City slickers can spend $100,000 for a frisky Holsteiner or $500 for a gentle warm blood mare and keep the animal in a stable. It isn't cheap. Horses eat half a bale of hay a day and need medical care. But to ride is to feel like Shakespeare's French prince who said of his steed, "When I bestride him, I soar, I am a hawk; he trots the air; the earth sings when he touches it; the basest horn of his hoof is more musical than the pipe of Hermes." Let's not get too carried away. It's just a horse.

912. The American Dream House: *Everyman's castle.* A 32-inch or bigger high def TV, a recliner, a pool—aboveground is fine—a king-size bed and a barbecue grill it took 12 hours to assemble. If you have these items, no matter how poor you are, dawg, you have arrived. Welcome to the good life.

911. 911: *The number to know when you need help.* We like the 2006 story about the beagle named Belle who saved her owner by using her teeth to dial 911 when her master lapsed into a diabetic coma. Sure, she'd been trained to do it, but the fact that she was actually able to pull it off in a life/death situation is worth celebrating. About 240 million calls for help are logged annually.

910. Adirondack Great Camps: *Woodland elegance.* Life is rustic but good in Great Camp style where multibuilding complexes of log cabins and lodges built with local timber and stone are filled with twig- and branch-fashioned furnishings. Country retreats for the wealthy generations years ago in upstate New York, these vacation spots still line the shores of Adirondack lakes. Our favored getaway: The century-old Hedges, where kayaking on crystal-clear Blue Mountain Lake is a joy.

909. The AAA: *Don't leave home without it.* Highway breakdowns don't have to be traumatic, not if you're one of the 50 million members of Triple-A, formerly known as the American Automobile Association. A phone call to this nonprofit automobile lobby group and service organization will bring you roadside assistance, its best-known benefit. When you're stuck on an interstate on a stormy night, it's a godsend.

908. Right on Red: *Panacea for the impatient.* Nothing worse than cooling your heels at an interminable red light when no traffic is coming in the other direction. The right-turn on red became a national trend in the 1970s as a gas-saving measure. Now, it's the law of the land, unless otherwise posted. Visitors to the Big Apple please note: it's a no-no in New York City.

907. Self-Help Books: *Striving for personal perfection.* Ever since Dale Carnegie taught us *How to Win Friends and Influence People* in 1936, we have been struggling to make ourselves better through books. We like *Why Is It Always About You? The 7 Deadly Sins of Narcissism* and the foundational *How to Read a Book*.

906. Boys and Girls Clubs of America: *Escape from the streets.* The goal is the same as in 1860 when the first club of its kind opened in Hartford, in 1931, when the federation of Boys Club chapters went national, and in 1990 when girls were added to the title: give children a safe place to learn, grow, and play under the supervision of caring adults. Michael Jordan, Jennifer Lopez, and two Bills—Cosby and Clinton—are alumni of the organization that now serves 4.5 million kids.

905. Baggy Bathing Trunks: *They suit us.* We believe in freedom of expression, but when it comes to skintight skimpy bikini bottoms favored by most of the world's men, we shout, "Too much information." While we're on the subject, we're glad the idea of topless bathing for women has flopped at most American beaches.

904. Mackinac Island: *Different emissions.* Personal motorized vehicles are banned from Michigan's picturesque 4.37-square-mile island in Lake Huron, hitherto a sacred Native American gathering place, fur trading center, then haven for the wealthy. So you walk, bike, or take horse-drawn carriages. In fact, Mackinac's 6,000 horses always have the right of way, providing steady work for manure collection teams.

903. Single-Sex Bathrooms: *To each, his or her own.* One of us will never forget how she spent a frustrating 15 minutes trapped in a stall in a unisex Parisian restroom as droves of Frenchmen continued to use the urinals right outside her door. Call us prudes, but we prefer to tinkle—er, mingle—with our kind.

902. Summer Jobs for Kids: *Keeping them occupied.* We'd like to thank the town recreation departments that paid our teenagers minimum wage during the summer months and kept boredom at bay. These part-time job opportunities have shriveled during tough economic times, but smart municipal governing bodies and federal

authorities recognize that keeping teenagers active and involved promotes good citizenship and civic engagement and is just as important as the senior citizen programs at the other end of the life spectrum.

901. Gown Towns: *Great for young and old.*

It's no coincidence that so many people retire to college towns. Not only do settings like Chapel Hill, North Carolina; Newark, Delaware; and the two Athens—Georgia and Ohio—bustle with the energy young adults can generate, but they have a certain intellectual cachet. They also provide downtowns alive with shops, eateries, and nightlife while serving as a regular source of inexpensive concerts, theater, and other cultural events.

900. Elite Country Clubs: *Lifestyles of the rich and, uh, rich.*

In these lush green places people in upper income brackets play golf, tennis, and cards, drink cocktails, swim in pools, eat pretty good food, and marry off their children. Isn't it great that there are still havens in America where people who have lots of money can congregate away from the lesser folks?

899. Acadia National Park: *Bah Hahbah's playground.*

No matter how you pronounce it, Bar Harbor, Maine, is one of New England's busiest tourist destinations, and the 47,000 acres of the-nearby park are among the prettiest on the eastern seaboard. The tide pools along the rocky shore are minimuseums of coastal critters.

898. Jeeps: *Legendary cruisers.*

These fabled off-road vehicles have caromed through countless war and adventure movies. Intrepid Indiana Jones drives one, naturally. So do unabashed romantics. Chrysler-made for more than 70 years, the four-wheel-drive Jeep can overcome anything from deep mud and snow to rugged terrain. But we humbly remind spirited Jeep drivers to refrain from ripping up the landscape.

897. Picket Fences: *Minimalist borders.* Especially popular in the United States, these are welcoming sights that can enhance yards, set off pretty gardens, and mark property lines. Invariably white, the fences are short, with tapered or pointed tops on evenly spaced vertical boards, called pickets. Wasn't this the sort of fence that Tom Sawyer conned his friends into whitewashing for him?

896. Kidspeak: *Word up!* If you are over age 30, this entry is sketch. Lol! Our young people always adopt a special language that excludes their parents. Whether your cell phone is blowin' up, you're out of cheddar, or that home run was filthy, it's mad good.

895. Lost and Found Departments: *Eureka! We found it!* Sometimes no more than a cardboard box under a receptionist's desk, these gold mines of mislaid objects speak to our innate honesty. If it ain't yours, you can't keep it. If it hadn't been for an active lost and found table at our local elementary school, we would have had to purchase about six coats one winter. Regular visits to the table, piled high with clothing and equipment, not only netted the aforementioned jacket, but provided a source of entertainment and amusement. Just how does a toilet-trained student lose underwear during the course of a school day?

894. Brooklyn Beer: *Drink this, ya bum.* What started in 1987 as a risky venture for former AP foreign correspondent Steve Hindy and ex-banker Tom Potter serves up some of the best suds in the country from breweries in Brooklyn and Utica, New York. It reminds us of the days when New York's most populous borough was synonymous with rude urban democracy. The company's sales volume has cracked the top 40 list among U.S. producers. When the weather gets hot, try a Brooklyn Summer Ale. On tap if you can find it.

893. Backyard Pools: *Affluence on the cheap.* For as little as $1,000, homeowners have cool places to relax, sip a drink, and make their kids the most popular ones on the block.

892. Sourdough Bread: *The other San Francisco treat.* Using a starter dough containing a lactobacillus culture, this form of bread-making was once common in Europe but is little used there now. Not the case in the United States. The bread was so popular during the California Gold Rush that "sourdough" became a nickname for a gold prospector. Our most famous version, a distinctly tart white bread, hails from San Francisco. The staple is so connected with that city's culture that "Sourdough Sam" is the mascot of the San Francisco 49ers.

891. Hayrides: *Hayseed heaven.* You're from the city and you want your kids to appreciate Nature. What better way than a hay-ride in crisp, cool autumn air? Climb on the truck bed stacked with hay and hunker down. The tractor pulling you will wend its way through farmland where there are pumpkins to be picked.

890. *Antiques Roadshow*: *Identifying our treasures.* For more than a decade, this popular PBS show has been educating viewers about the value of family heirlooms and our attics' dust collectors. Who knew? That tattered Navaho blanket you've had tossed over a chair is worth $350,000. That dirty card table you picked up at a garage sale for $25 is really worth $300,000, and great-grandpa's ceremonial sword is actually valued at $200,000. It's enough to make you wary of ever throwing anything out.

889. Free Refills: *Drink up—there's more where that came from.* We return from a trip abroad renewed in our love of this oh-so-American custom. Diners and restaurants around the United States keep our soda glasses full and top off our coffee cups, but only charge once for the service.

888. ASPCA: *Animal defenders.* The American Society for the Prevention of Cruelty to Animals first protected New York City's working horses in 1866. It was the first humane organization in the Western Hemisphere. Now, it can investigate cruelty to all animals and even make arrests. Other services: adoption centers, a poison control center, animal training and medical care, even grief counseling when Spot dies.

887. Cuyahoga Valley National Park: *Escape the crush.* Ohio ranks ninth among states in population density, so it needs as much open space as it can get. The only national park is in the ideal spot, the middle of the Cleveland-Akron megalopolis. Beginning bird watchers will want to join an organized expedition to spot some of the park's 200 species.

886. Buses for Old Folks: *The senior service.* For old people too shaky to drive or too poor to own a car, free transportation to the store or doctor is a lifeline. In many places, like Kokomo, Indiana, a bus shows up at the house of anyone over 60 who calls the day before and wants a ride. We know a senior on a pension who takes such a bus every day. He's the driver.

885. *Moonstruck*: *Glowing romantic comedy.* Wacky, warm, and wise says it all for this 1987 Norman Jewison film which isn't high on "best" lists, but we cherish it for its inventive script, endearing characters, and on-the-mark acting. The lovers are a widowed Brooklyn bookkeeper who lands in bed with her fiancé's brother, a one-handed, misanthropic baker. But then, anything can happen with a big fat moon overhead. Our favorite scene: the bookkeeper (played by Cher) slaps the whining baker (Nicolas Cage) across the chops, crying, "Get over it!"

884. Craters of the Moon National Monument and Preserve: *Mighty magma.* Beginning over 15,000 years ago, lava eruptions have shaped this volcanic landscape on the Snake River Plain in southern Idaho. The area's three major lava fields and sagebrush grasslands cover 1,117 square miles. So lunarlike is the terrain that NASA astronauts explored the monument in 1969 while training to visit the moon.

883. Legacy of Watergate: *A healthy skepticism of politicians.*
For nearly two centuries, the American press quietly kept political secrets from the public. Then along came *Washington Post* reporters Carl Bernstein and Bob Woodward and their dogged pursuit of the real culprits behind a burglary of the Democratic National Committee office in the Watergate building in Washington, D.C. The journalists' persistence, coupled with the steadfast pursuit of truth by U.S. District Court Judge John Sirica, led straight to the top, ultimately toppling the administration of President Richard Nixon in 1974. It also forever changed the American attitude toward our politicians—we know they lie. Our inclination now is to believe the whisperings we hear about extramarital affairs, blackmail, influence-peddling, and other unseemly ethical and legal no-nos.

882. Table Browsing: *I'll have what they're having.* Walk
through an unfamiliar restaurant and keep your eye on the plates on other tables. Decide what looks best and order that. It's a great way to try food you might never have thought to order. "I'll have what they're having" works every time.

881. Baltimore Inner Harbor: *Model for a turnaround.* Mary-
land's largest city seemed to be a down-at-the-heels urban nowhere until the 1980s, when redevelopers turned the harborside into one of the best attractions on the East Coast. Each year, millions of tourists visit the old ships, museums, night life, restaurants, history sites like Fort McHenry, and big-time sports at nearby Oriole Park at Camden Yards and Raven's Stadium.

880. Notre Dame Football: *Millions cheer her name.* Of all college teams, the Fighting Irish of South Bend, Indiana, has the best history, the best fight song, the most fans in states other than its own (that's a guess), and the best movies *Rudy* and *Knute Rockne All American*. In that film, Ronald Reagan spoke lines replayed almost as often as "Mr. Gorbachev, tear down this wall." He (Reagan, not Gorbachev) played star George Gipp on his deathbed, whispering to the coach, "Some time, Rock, when the team is up against it, when things are wrong and the breaks are beating the boys, ask them to go in there with all they've got and win just one for the Gipper." We spotted the happy ghost of Gipp on a frosty fall Saturday lurking among 80,000 fans at Notre Dame Stadium, some wearing green paint and not much more.

879. Alinea: *Cutting-edge cuisine.* Most of us cherish beautifully prepared meals. But the experience of eating at Chicago's Alinea restaurant is rapturous because chef/owner Grant Achatz's creativity, and his penchant for deconstructing classics, boggles the taste buds. Two examples: the Black Truffle Explosion with black truffle liquid, romaine, truffle, and Parmesan, and Hot Potato, Cold Potato, a soup laced with butter, chives, cheese, and, again, truffle. A tasting menu costs $145, but it's well worth it, considering that in 2009 *Restaurant* magazine pegged Alinea as one of the best in the world.

878. *The Deer Hunter*: *War redux.* On one level, director Michael Cimino's 1978 epic with its all-star cast shows the impact of the Vietnam War on hometown America. Digging deeper, it's a story of friendship, loyalty, and, ultimately, the sanctity of life. The plot involves three steelworker friends who go to fight in Vietnam and are captured and tortured by the Vietcong. Two come back, one in a wheelchair, the other a troubled hero. The third stays behind, so demented that he plays Russian roulette—the game his Vietnamese captors forced on him—for high stakes in Saigon.

877. Palmer House: *Chicago grandeur.* In its third incarnation since 1871, the huge Palmer House hotel stands 25 stories proud at the corner of State and Monroe streets. Now known as the Palmer House Hilton, the dowager recently underwent a $170 million face-lift. Elegant once more, the hostelry has even revived its stretch of the Loop.

876. Fresh Baked Goods: *Second only to France.* We love coffee cake for breakfast, a bagel with cream cheese for lunch, and crispy garlic bread as a complement to our dinner. The best rolls you have ever eaten can be found at Brothers Bakery in Kearny, New Jersey. Take exit 15W off the New Jersey Turnpike and follow your nose (and your GPS) to this nondescript storefront where you will find "hard rolls" (they call them Kaiser rolls elsewhere in the U.S.) that are slightly crisp on the outside, but soft and delicious inside. The ones with poppy seeds are the tastiest. You can buy baguettes, rye, and country white round loaves of bread. We also recommend the brownies.

875. Warped Tour: *Summer grunge to go.* This traveling festival of punk, ska, and other out-of-the-mainstream genres gathers dozens of promising musicians together with a few established bands like NOFX and Bouncing Souls and showcases their music around the country. After 15 years, Warped Tour is becoming part of the musical establishment, but we still like its edginess.

874. Olympic National Park: *Wilderness in Washington.* Its 933,000 acres provide a triple play of wonders—forests, mountains, and a pristine coastline 73 miles long. See the world's largest unmanaged herd of Roosevelt elk. And take rain gear. The park's Hoh Rain Forest gets up to 14 feet of precipitation each year.

873. Traditional Marriage Proposals: *How to say "Let's do it."* Popping the question in front of 50,000 fans at a ballpark became passé when companies started staging such stuff for a fee. Holding a "Marry Me" sign as you skydive past your lover? Hot stuff today, but easily topped tomorrow when some idiot does it without a parachute. C'mon, young America. Buy a ring. Drop to your knee in a park or beach or on mountain peak at sunset. Look your loved one in the eyes and say something romantic, like, "Will you share your life with me as my wedded whatever?"

872. T Rex Returns: *Monster fantasy.* The recent astonishing discoveries of soft tissue in fossilized dinosaur bones—including blood vessels and cells in a Tyrannosaurus rex unearthed in Montana, and organic molecules in the skin and tendon of a hadrosaur in North Dakota—have revived dreams of creating a real Jurassic Park. As excited as scientists are about the finds' research potential, however, they doubt they'll be able to isolate dinosaur DNA and bring the creatures back to life. But you never know.

871. Mysterious Disappearances: *The ultimate national puzzle.* Writer Ambrose Bierce reportedly was last seen fighting for Pancho Villa in Mexico, aviator Amelia Earhart was lost over the Atlantic, hijacker "D. B. Cooper" parachuted into a forest with ransom loot, union boss Jimmy Hoffa went to an unmarked grave, and the entire Roanoke colony (including Virginia Dare) vanished. And what about all those sailors and fliers who met their end in the Bermuda Triangle?

870. Clambakes: *Beach blasts.* True to Native American practice, harvest fresh seaweed brimming with saltwater bubbles, the source of steam for cooking. Build a four-foot-deep pit in the sand, line it with charcoal and rocks, then ignite the lot. Once the fire settles,

pile on whole lobsters, steamers and mussels, corn on the cob, and red-skinned potatoes. In no time it's a seaside feast.

869. The Natchez Trace Parkway: *One long and winding road.* No billboards, malls, or truck stops mar the beauty of this 444-mile "greenway" that runs past forests from Natchez, Mississippi, through Alabama to just south of Nashville, Tennessee. Once a foot-path north for traders, it now draws campers, hikers, and horse riders. Best bet: the magnificent views at Leiper's Fork.

868. The Two Best Days in a Boat Owner's Life: *Hello and goodbye.* About 8 million boats are registered in the United States, ranging from little putt-putts to the $100 million yachts with helipads favored by business moguls. Donald Trump had his first best day when he bought a 110-foot cruiser and called it the *Trump Princess.* He had his second best day in 1991—and so did the creditors of his gambling empire—when he sold the @$*&# thing to a Saudi Prince for $40 million.

867. College Team Nicknames: *Lions and Tigers and Bears and Salukis.* Being true to their schools, the authors of this book cheer for their alma maters' teams, the Pirates, the Seawolves—who used to be the Patriots—the Violets, whose mascot is a Bobcat, and the curiously contradictory Fighting Quakers. Some colleges don't want to emulate savage beasts and choose monikers like the Fighting Artichokes of Scottsdale Community College and the Banana Slugs of UC Santa Cruz. In 1995 a Tiger building on work by a Golden Bear solved Fermat's Last Theorem. Next big riddle: is Georgetown's Hoya an animal, mineral, or vegetable?

866. Televangelist Meltdowns: *The gods that flailed.* The list of miscreants is so long and the sordid details of their exploits so well known (even the *New York Times* covers them in the guise of exploring the nation's social mores) that we can make this entry brief. Just close your eyes and think about all those pompous wind-bags forced to make teary-eyed confessions in front of TV cameras. Sweet!

865. Indiana Jones: *Down-to-earth superhero.* A tough guy who's afraid of snakes—who can resist that combination? A professor of archeology who wields a bullwhip—what's not to like? The globe-trotting adventurer created by George Lucas and brought to life on screen in four films directed by Steven Spielberg always has a ready quip as he plunges into fights to the death with Nazis and other bloodthirsty adversaries.

864. Theodore Roosevelt National Park: *Honoring a visionary.* The future chief executive first traveled to North Dakota's Badlands to hunt bison in 1883 and ease his grief over the deaths of his wife and mother. The visit was a life-changing event. "I never would have been President if it had not been for my experiences in North Dakota," he said. This vanishing frontier shaped Roosevelt's attitudes about protecting our natural resources. Today, the 110-acre park that bears his name contains several sites along the Little Missouri River, including a petrified forest, Wind Canyon, eroded badlands, herds of wild horses, and Roosevelt's Elkhorn Ranch cabin.

863. Toys For Tots: *Even tough guys have heart.* This program began in 1947 when Major Bill Hendricks and fellow Marine Reservists in Los Angeles distributed 5,000 Christmas toys to needy children. That effort was so successful that the Marine Corps adopted Toys for Tots and expanded it into a nationwide campaign. Since then, Toys for Tots and its namesake foundation have distributed more than 440 million toys to more than 173 million children.

862. The Big Broadcast: *Keeping our grandparents' music alive.* Sunday night, eight to midnight, for most of the past 35 years, record-hunter Rich Conaty has been playing jazz and popular music from the 1920s and 1930s on commercial free WFUV-FM in New York. Now the show is on the Internet. Mildred, Bix, Ukelele Ike, Bing, Bessie, Fats, Duke, and Cab will make you happy.

861. Franklin Institute: *Science in action.* We love this Philadelphia science museum's overnight adventures for preteens. Long before Ben Stiller spent a *Night at the Museum* in his comedy film, the Franklin Institute was inviting student groups to sleep in and experience the exhibits in a whole new way. We salute such initiatives that teach our youth the enrichment that comes with a visit to a museum.

860. Creative Criminal Sentences: *When punishment fits the crime.* A judge in Florida sentenced teen traffic offenders to keep a scrapbook of newspaper clippings about auto accidents. Two Ohio men who taunted a woman and threw beer bottles at her car were ordered to dress in women's clothing and walk down Main Street. A judge in Texas made a wife-beater learn to chill out at yoga classes. And in Vermont, a judge followed the state's suggestion and ordered two dozen teenagers who vandalized Robert Frost's summer cottage-turned-museum to take a Middlebury College course on the poet's life and work. "I thought that if these kids learned something about Robert Frost and his life they'd have more respect for Frost in the future," the prosecutor said.

859. The PGA Tour: *Where par is subpar.* The best golfers in the world amaze us every week at Professional Golfers Association tournaments by taking just two to four shots to get the ball into a hole sometimes as far as 600 yards away. Three golfers hold the record low score of 59 for an 18-hole course in PGA events.

In golf, the opponent is oneself. And a plastic sphere 1.68 inches in diameter. "It will always be the ball and me," said Tiger Woods, the dominating golfer whose assault on Jack Nicklaus's record of 18 victories in major championships was put on hold by a 2009 sex scandal.

858. Bottle Bills: *Litterbug busters.* Eleven states have laws that require a deposit of a nickel or a dime on beverage containers, and campaigns are afoot in several others to enact similar measures. Why target bottles and cans? Because they account for half of all litter. Some proponents—including us—argue the deposit should be bumped up to a quarter to account for inflation since the first bottle bill was enacted in Oregon in 1971. If that happened, America's roads, parks, and hiking trails would become pristine overnight.

857. The National Do-Not-Call Registry: *Peace at Last.* We no longer have to listen to those annoying dinnertime phone pitches. This government-enforced system for banishing telemarketers from calling us at home if we have placed our numbers on the list is why. (Cell phones have always been excluded from solicitations). Loopholes exist: charities, pollsters, and political groups, for example, are not bound by the law. Nevertheless, things have gotten a lot quieter since it went into effect in 2003.

856. The Tournament of Roses: *Floral fantasia.* Pasadena, California's Rose Parade has been a New Year's Day tradition for more than a century. Millions of flowers go into decorating the colorful floats. It's all a showy prelude to the Rose Bowl Game, the annual college football joust nicknamed "The Granddaddy of Them All" because it's the oldest bowl game in the country.

855. Man Caves: *The better half's hangout.* Guys need a refuge from chintz, damask, and lace. A space where they can knock back some beers, surround themselves with leather and stonework, smoke cigars, and watch sports on mammoth, flat-screen TVs. So go convert that unfinished basement into your own cave and be a manly man. Need ideas? Visit the Official Man Cave Web site.

854. Muffuletta: *More than a sandwich.* When you go to New Orleans, head to the Central Grocery in the French Quarter and order one of their indescribable muffulettas. This thick combination of deli flavors on a round Italian roll can be found at other restaurants around the city, but the Central Grocery is where it was first created in 1906 and where it is still the best.

853. Romantic Comedies: *Where the guy eventually gets the girl.* From *It Happened One Night* in 1934 to 1972's *What's Up, Doc?* to *He's Just Not That into You* from 2009, we love this movie genre. Predictable? Absolutely, but there's something very appealing to sitting down with a mindless comedy and a lovely romance for a few hours. Sometimes called "chick flicks," these movies celebrate the rocky, unpredictable, and sometimes hilarious road to true love. Our favorites? *The Philadelphia Story, An American President*, and *Dave*.

852. Car Detectives: *Spotting the clunkers.* Carfax is the best known of the companies that sells reports based on car ID numbers that tell whether the auto you plan to buy has been in a crash. Or whether the seller is lying when he says the vehicle's only owner was an old lady who drove it only on Sundays to church. The U.S. Justice Department is getting into the act with a similar service, the National Motor Vehicle Title Information System Resource Center, or NMVTISRC for short. Catchy, huh?

851. Weird Footwear for Kids: *Small-fry fashion statements.* From slippers shaped like ducks to shoes that flash colors or double as roller skates, these prepubescent foot fetishes always make us smile. Some are worth buying just for the name, like the glow-in-the-dark Nickelodeon Slimers Spazmodium we saw on sale for $29.99.

850. Legends of Lost Treasures: *The ultimate get-rich-quick dream.* From Pegleg Smith's Lost Mine in California to buried pirate hoards in Florida, stories of lost caches of gold and gems excite our greed as nothing else. Christ knew this, saying: "The kingdom of heaven is like a treasure hidden in a field" that a man finds and buys from the clueless owner. Hidden booty seems to be everywhere in this country; a rancher we know claims an old cowhand stashed a sack of riches *somewhere* on his spread—but where?

849. Amber Alert: *Kidnappers' nemesis.* AMBER plans are voluntary partnerships between law enforcement and public broadcasters to send out messages over the Emergency Alert System notifying communities that a child has been abducted and asking them to aid police in recovering the youngster. Created in 1996, the program operates in all 50 states. By 2009 it was credited with saving the lives of more than 440 children. Dubbed "America's Missing: Broadcast Emergency Response," the effort honors nine-year-old Amber Hagerman, who was kidnapped while riding her bicycle in Arlington, Texas, and then murdered.

848. Guide-Dog Programs: *Canine companions.* An estimated 5 million Americans age 65 and older are blind or severely visually impaired. This number could more than double by the year 2030. Despite this grim outlook, there's hope for these disabled—and younger ones—from highly trained guide dogs that give the blind better mobility. To this end we salute the Guide Dog Foundation for the Blind, Inc., Smithtown, New York, a venerable national nonprofit whose free services include a guide dog, four weeks of training in its use and care, board and lodging for the student at the training center, and transportation costs.

847. Chinese Takeout: *Yum yum dim sum.* When we lived in Europe in the 1980s, a friend used to stop in Chinatown en route to New York's Kennedy Airport. She'd pack fried dumplings and prawns with walnuts in an insulated pack and whisk them off to Brussels for the ultimate take-out feast. Easy access to excellent Chinese takeout is one of the reasons to live in the New York and San Francisco metropolitan areas. Our favorite: an order of sautéed string beans over rice.

846. Action/Adventure Movies: *Keeping us on the edge of our seats.* Who doesn't love Harrison Ford hurling a bad guy into oblivion with the line, "Get off of my plane!"? Or what about Arnold Schwarzenegger as the Terminator promising, "I'll be back"? The action/adventure genre is a Hollywood staple loved by movie executives because it brings in the guys for the shoot-'em-up fun and the girls for the romantic heroes.

845. Mama E's Waffles and Wings: *The soul of the Southwest.* Millions of diners have a favorite soul food joint. Keith Patterson's shack of a restaurant in Oklahoma City deserves consideration for no. 1. Patterson named the place after his mother and his most famous offering: a plate of flaky waffles topped by crispy chicken wings smothered in powdered sugar and maple syrup.

844. The Cleveland Orchestra: *Cultural oasis on the Erie.* For a city with fewer than 500,000 people to have one of the world's greatest symphony ensembles is a tribute to Cleveland's reverence for sounds that soothe savage breasts. The city has supported the orchestra since its start in 1918. Autocratic maestro George Szell, who ruled from 1946 to 1970, created the reputation, and subsequent conductors like Franz Welser-Most have kept the tradition alive in Severance Hall.

843. Rube Goldberg Machines: *Comically convoluted contraptions.* Goldberg's famous cartoons of zany devices have entered the national lexicon to describe any pointlessly complicated operation. In the "Local Government Efficiency Machine," a taxpayer sits on a cushion, forcing air through a tube to blow a balloon over a candle, with the resultant explosion scaring a dog into pulling a leash attached to a board that tilts to release a ball . . . Well, you get the idea. After many more steps, an artificial hand powered by the growth of a house plant removes the taxpayer's wallet from his back pocket. Goldberg, who died in 1987 at age 87, won the Pulitzer Prize in 1948 for his wacky renditions.

842. Teach For America: *Lessons learned.* Princeton alumna Wendy Kopp created this non-profit program in 1990 to encourage outstanding college graduates and professionals of all academic majors and career interests to become teachers in hard-pressed urban and rural areas. Since then, this movement to eliminate educational inequity has enlisted more than 24,000 instructors to reach over 3 million students.

841. The Ten Most Wanted Fugitives: *Infamous list.* Since the FBI began publicizing these descriptions in 1950, 462 of the 491 individuals named have been located, about a third of them thanks to public help. Lately, the Bureau has used *America's Most Wanted: America Fights Back*, hosted by John Walsh on Fox Television Network, to solve crimes and snare dangerous criminals. A nationwide toll-free hotline allows viewers with information about any of the criminals to provide anonymous tips to law enforcement officials involved in the investigations.

840. She Crab Soup: *Southern comfort.* Boston and Manhattan have their chowders, but Charleston, South Carolina, has She Crab Soup. The devotion to this regional favorite is complicated by the fact that it is illegal to trap mature female crabs. Thus, many chefs rely on the more plentiful males. The soup is standard crab bisque dressed up with a splash of yellow crab roe. Since the oddly colored eggs aren't for everyone, the recipe has evolved to include a garnish of chopped eggs instead.

839. *The Godfather*: *Romanticizing bad guys.* Our obsession with the Mafia can be traced back to Don Corleone, the Marlon Brando character in Mario Puzo's best seller turned 1972 blockbuster film sensation. Directed by Francis Ford Coppola, the movie had two sequels. It introduced phrases like "swim with the fishes" into our vocabulary and helped us imagine the inner workings of this secretive crime family. What *The Godfather* did for the Mafia in the 1970s, the HBO series *The Sopranos* did for the family in the 1990s. If you haven't seen them, watch them now. It's an offer you can't refuse.

838. Alternative Dispute Resolution: *Saving the expense of trials.* Whether it's an argument between neighbors over a noisy dog or a multi-billion-dollar row between Fortune 100 companies, the savvy way to resolve civil complaints is mediation and arbitration, not a trial. So far, relatively few cases are settled by such means. A 2006 study in New York State, for instance, found that only 36,000 of the 4.5 million trial court filings escaped the full-blown litigation that lawyers love. But many states now make attempts at mediation mandatory for such cases as divorces and landlord-tenant disputes. And there's always TV's *The People's Court* and *Judge Judy* for litigants who want to go to court without being in court.

837. Zagat's Guides: *Word of mouthfuls.* In 1979, Nina and Tim Zagat started compiling their friends' ratings of New York eateries. Now their books based on thousands of customers' evaluations of restaurants, hotels, and activities in more than 100 locations are among America's most popular guides to the good life.

836. All-You-Can-Eat Buffets: *Windfall for gluttons.* They gotta eat too, and the impecunious among them can turn to these culinary cornucopias to bulk up for lean days ahead. If you have a fatty-food fixation, fixed prices are your ticket to low-cost Nirvana. We heard of a place in Harlem that served all-you-can-eat dinners for $11.95. It closed. We wonder why.

835. Graduated Driving Licenses: *A safer route.* Also known as provisional licenses, this system phases in young beginners to full-driver status based on maturity and driving aptitude. Florida instituted it in 1996. Now most states apply it. Ideally, the plan has a learner's stage beginning at age 16 and lasting at least six months, 30 or more hours of supervised driving, plus restrictions on unsupervised night driving and passengers. Teens may grumble, but parents applaud it. States with these elements on their books show teen crash reductions of from 10 to 30 percent.

834. Moore's Law in Action: *Miracle of memory.* The proposition, which says the number of transistors that can be crammed on a chip doubles every 18 to 24 months, is a tribute to the brilliance of U.S. manufacturers who produce amazingly powerful computers, gadgets like smart phones and digital cameras, and life-saving medical-diagnostics technology. Though this shrinkage is approaching its physical limit, we're confident our nanotechnology scientists will find the way to even greater switching power. Intel cofounder Gordon E. Moore came up with the idea in 1965.

833. *Taps* on a Bugle: *Mournful, yet beautiful.* This haunting 24-note melody began as a Civil War bugle call that signaled lights out and that all was well. Today it sounds at U.S. military funerals, memorial services, and wreath-laying ceremonies, and to signal the end of the meeting for Boy and Girl Scouts alike.

832. The Brotherhood of Bald People: *Down with comb-overs!* This Coon Rapids, Minnesota, organization helps the world's 1.4 billion bald people—especially men suffering from male-pattern baldness—accept their shiny pates. We applaud their efforts because one of the worst things about America is men who pretend they have a full head of hair by stretching their part halfway down their head and combing the hair over. Donald Trump, are you listening? You're bald!

831. American Wildflowers: *Horticultural heritage.* Who knew that invasive plants like burning bush or Japanese barberry threaten our wildflowers? Not to worry. Garden in the Woods, Framingham, Massachusetts, for one, preserves and nurtures more than 1,000 native plant species, including the rare, endangered Plymouth gentians and coral-root orchids. These and other flowers populated forests, marshes, prairies, and mountains long before the Pilgrims arrived.

830. *Moby-Dick*: *A whale of an epic.* Herman Melville's 1851 work is considered one of the greatest novels in the English language. This story of a wandering sailor, his voyage on the doomed ship *Pequod,* and its obsessed captain, Ahab, has everything: symbolism, biblical references, adventure, tracts on the natural world, and themes of idealism versus pragmatism, revenge, and politics. The book also delivers the most famous opening line in American literature: "Call me Ishmael."

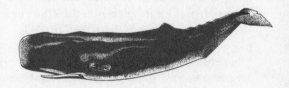

829. Whale Watching: *Thar she blows.* Boats cruise from ports like Seattle, San Diego, and Honolulu to observe these behemoths. Ships off Provincetown, Massachusetts, for example, regularly record a procession of humpbacks, minkes, and fins. But you needn't sail to see them. Annually, thousands of gray whales swim close to Oregon's rugged coast on their 12,000-mile migration from the Arctic to Mexico and back.

828. A Prairie Home Companion: *Way-above-average broadcast.* This National Public Radio hit attracts some 4 million listeners every Saturday to hear Garrison Keillor and company present wacky spoofs, tall tales, and country music. Regular bits on this show that began in 1974 include "Guy Noir (Private Eye)," "Dusty and Lefty," "The Guys' All-Star Shoe Band," and bogus ads for Powdermilk Biscuits and the Duct Tape Council. Then there's always "News From Lake Wobegon," the fictional Minnesota town where "all the women are strong, all the men are good-looking, and all the children are above average."

827. McDonald's French Fries: *Tasty grease.* These fries make this list, but only if they are fresh from the fryer and covered in salt. Five minutes later the thrill is gone. The treat comes at great cost, however. One large order of fries packs a third of the day's recommended calories.

826. Wal-Mart: *Haven in tough times.* When unemployment rises, Wal-Mart stock does too. The world's biggest retailer gets plenty of grief for its gobble-up-the-little-guy philosophy, but it's hard to ignore a 30-inch flat-screen HDTV for less than $350. Employer of more than 2 million people, Wal-Mart slashes prices to bring customers in the door. We loved the 1995 Billie Letts novel *Where the Heart Is*, about the abandoned pregnant teenager who secretly lived quite comfortably in a Wal-Mart when she had nowhere else to go.

825. National Enquirer: *What inquiring minds want to know.* We know it's trashy, but we can't help perusing the headlines while we wait in line at the supermarket.

824. Irish Coffee: *Caffeine with kick.* Think the Irish invented this tasty combination of coffee, whiskey, sugar, and cream? Think again. Irish coffee was first served at the Buena Vista bar in San Francisco in 1952. It's still the best place to go for one—bartenders there mix about 2,000 Irish coffees a day.

823. Play-Doh: *Rainy day fun.* We love this stuff. Introduced in 1956, Play-Doh, which behaves like nontoxic modeling clay but doesn't harden quickly, provides hours of fun for preschoolers who want to roll, squish it, and, when Mom's not looking, eat it.

822. Baseball Cards: *A childhood treasure beneath beds and inside attic storage bins.* Baseball cards are only valuable because so many moms threw them away. Now, Moms know better and they cling to dozens of boxes filled with near worthless rookie cards for guys who almost, but not quite, made it in the big leagues. Yet we celebrate this nostalgic remnant of childhoods' past. You also can figure out someone's age by whether they remember the thin strips of bubblegum tucked in a pack of baseball cards. Topps took the gum out of the package in 1991 because collectors complained that it was staining their mint condition cards.

821. Saint Cupcake: *A little bit of heaven.* The best way to eat a cupcake is to pull the cake part in half, then put that piece on top of the icing and eat the cupcake like a sandwich. Delish. We recommend you stop by Saint Cupcake in Portland, Oregon, on a Saturday, when the bakers make their Red Velvet and Carrot Cake cupcake specials.

820. Do Overs: *The childlike way to get it right the second time.* Got a disputed foul ball? A tennis serve that might have been out?

Dice that fall on the edge of the board game? American kids have the answer: "Do over! Do over!" yelled loudly and passionately. If only all of life were so easy.

819. U.S. Women's National Soccer Team: *Kicking butt.* The team hails from a country that treats soccer—or football, if you listen to the rest of the world—like a poor stepchild, yet the U.S. women's team has brought home unprecedented back-to-back gold in the last two Olympics. Looking for role models for your preteen daughter? Take her to one of these soccer matches for inspiration.

818. Mr. Met: *Sport mascot with a big head.* When he debuted in 1964, Mr. Met was Major League baseball's first live-action mascot. With his larger-than-life hardball head and his ever happy countenance, he's still the best. We grudgingly acknowledge Mr. Met's nearest rival, the energetic Phillie Phanatic, who might have finished first if only he rooted for another team.

817. Church Signs: *The word of God?* Maybe God has a sense of humor. Consider these messages from church signs around the country: DO NOT WAIT FOR THE HEARSE TO TAKE YOU TO CHURCH; THE 10 COMMANDMENTS WERE NOT CALLED THE 10 SUGGESTIONS; FORGIVE YOUR ENEMIES: IT MESSES WITH THEIR HEADS; and CHRISTMAS: EASIER TO SPELL THAN HANUKKAH.

816. Alaskan Salmon: *Swimming upstream in the economic pipeline.* Commercial fishing of this popular fish is in the pink. What started in the 1880s as a canning industry now accommodates the public's taste for fresh wild salmon. The state harvests about 175 million annually, making it one of the most popular food fish in America.

815. Antique Malls: *Old stuff under one roof.* We used to wander the streets of quaint towns like New Hope, Pennsylvania, or Niwot, Colorado, in search of dusty antiques shops. But now we prefer the cooperatives that bring hundreds of dealers together under one roof. Check out the Memory Lane Antique Mall in Sevierville, Tennessee, the Williamsburg Antique Mall in Virginia, or the Heart of Ohio Antique Center in Springfield.

814. Supermarket Coupons: *Big business for conscientious clippers.* About 90 million people in the United States rely on coupons to save money. We salute people like the Coupon 101 instructor Tanya Senseney of Lady Lake, Florida, who paid $45 for her $457 shopping order. The cash register tape was 12 feet long!

813. Wyoming Dinosaur Center: *Welcome to Jurassic Park!* At this natural history exhibit, you can participate in a paleontological dig of the Jurassic rock where dinosaurs roamed 145 million years ago. Located in Thermopolis, the quarries on the Warm Spring Ranch regularly yield interesting dinosaur and fauna fossils.

812. X Games: *Extreme sports.* There are people who actually like to bungee jump, flip their dirt bikes, or catapult off the side of a wall atop a skateboard. For them, there are the X Games. These semiannual televised gatherings of the world's daredevils—in summer and winter—keep the roads and streets safer for those of us who would prefer to watch it all unfold from our couches.

811. Cow Chip Throwing Contests: *Hands-on fun at farm fairs.* Don't pooh-pooh this sport. What could be more fun than flinging a patty of dried bovine dung farther than the other guy? The folks in Beaver, Oklahoma, the self-proclaimed cow chip capital of the world, say the record toss is 182.3 feet.

810. Snowboards: *Another way to hurt yourself in the winter.* They are really surfboards for snow. Invented in the United States, snowboarding became an Olympic sport at the 2006 winter games. Novices might want to wear a wrist brace: for the record, broken wrists are the most common calamity.

809. World's Largest Ball of Twine: *All three of them.* Is it the one Frank Stoeber started in Cawker City, Kansas, in 1953 and is 40 feet wide and growing? Is it the one in Darwin, Minnesota, also 40 feet, that Francis Johnson built from 1950 to 1979? Or is the winner in Branson, Missouri, which was built with a machine? Visit all three and decide.

808. Weather Vanes: *Telling us which way the wind blows.* Weather vanes have been around just about forever, but in the United States they occupy a popular niche as folk art. While they performed a valuable farming service in years past, today's mostly decorative weather vanes retail for about $350. A 100-year-old Indian chief copper design netted $6 million in an auction at Sotheby's.

807. Going to the Mall: *Recreational browsing for teens.* The modern and safer suburban equivalent of the street corner, the mall is the place where children as young as middle-school age can meet friends without parents lurking and keep sales clerks at bay by saying, "Just looking."

806. **Advice Columns**: *Common sense answers to life's perplexing problems.* In newspapers, magazines, and now online, expert advice is just a page or a click away. What started as lovelorn columns in 18th century newspapers has transformed into syndicated big business. Afraid to ask your doctor about those unsightly warts? Pen an anonymous letter to Dr. Donohue and sign it, "Uneasy in Oshkosh."

805. **Cabela's**: *This chain huuunts.* For lovers of outdoor pursuits like hunting, fishing, camping, and boating, the 30 or so Cabela stores from Maine to Arizona are museums of gear. Besides the 85,000 square feet of space for products, most Cabela outlets have restaurants and classes. Try the bison bratwurst and then take fly-tying lessons.

804. **Newport Mansions**: *Summer playground for the 19th century's nouveaux riches.* When the robber barons wanted to escape their hectic city existence, they retreated to Newport, Rhode Island. There, they built expensive beach "cottages" that now stand as museums for us common folks. We vote The Breakers, the Vanderbilt family's 70-room bungalow, the most decadently gorgeous of all.

803. **#2 Pencils**: *Old reliables.* Let's throw away those cheap ballpoint pens and use good, old-fashioned, smooth-writing #2HB pencils sharpened to a fine point. The HB stands for "hard, black." The pink eraser at the end comes in handy. If you buy America's oldest brand, Dixon-Ticonderoga, you will be helping world trade. An Italian conglomerate bought the company in 2004.

802. Pastrami on Rye: *What's not to like?* Take beef that is corned, boiled, brined, and smoked. Add spices. Or skip all the work, go to the Carnegie Deli in Manhattan and enjoy.

801. The U.S. Resident Line at Passport Control: *There's no place like home.* You arrived exhausted on the flight from Prague or Jakarta or Lima, and it seemed like forever before the luggage came. Now the Homeland Security officer checks your passport, smacks it with a stamp, slides it toward you and says, "Welcome home." And you are glad you are back.

800. Rosie the Riveter: *She can do it!* This World War II symbol of women's influx into the workforce lives on. You can still buy a copy of the Westinghouse poster of a working woman rolling up her shirtsleeves under the slogan, "We can do it!" Now an instantly recognizable feminist image, Rosie's mug also can be purchased on a coffee mug, T-shirt, or book. Profits from the sale help the National Park Service tell the story of WWII on the home front at a park in Richmond, California.

799. Mathew Brady Photographs: *The Civil War in black and white.* Brady supervised a team of photographers whose pictures of humdrum camp life and the corpse-strewn aftermath of battles have given the world a visceral sense of the War Between the States. Congress bought his collection of 10,000 photographic plates for $25,000 in 1875.

798. Christmas Villages: *Twinkling nostalgia in miniature.*
What started with two buildings—a ceramic church and a tavern—
by our Christmas tree has grown into an elaborate layout that includes a Chinese restaurant, a university, and a fishing hole. We love
the sparkle of the tiny Christmas lights and imagine elaborate stories
about the secret lives of all 62 inhabitants. God bless them, everyone.

797. Fish Tacos: *Tropical twist on a Mexican favorite.* If you
want to make them yourself, track down Food Network chef Bobby
Flay's easy recipe on the Internet. We've been told on reliable authority that the fish tacos at Kona Tacos on the Big Island of Hawaii
are worth the airfare.

796. Little Golden Books: *The Poky Little Puppy and friends.*
Published in 1942, the Little Golden Books, with their distinct golden
binding, sold for 25 cents. That made book ownership possible for
children during tough economic times. These simple stories have become classics that are still a bargain at about $3. To date, more than
2 billion have sold. Our personal favorite: *The Golden Egg Book* by
Margaret Wise Brown, with *The Saggy Baggy Elephant* a close second.

795. Palo Duro Canyon: *Wild Texas habitat.* America's second
longest canyon is one of the few open spaces Texas has saved from
private developers. Lucky visitors to the 120-mile-long preserve will
spot the rare Texas horned toad.

794. Tiffany's: *Bling-bling at its best.* Baubles and beads, nestled in a Tiffany blue box, have wow power. The store, headquar-
tered on Fifth Avenue in New York City, provided swords for the
Union Army in the Civil War, designed White House china, crafted
Super Bowl and NASCAR trophies, and sold countless engagement
rings in the distinctive Tiffany solitaire setting.

793. The Georgia O'Keeffe Museum: *Flowers in the sand.* A pilgrimage to the largest repository of works by the artist of the American desert will take O'Keeffe lovers to Santa Fe. Among the 1,149 paintings, sculptures, and drawings are some of her famous floral pictures. The museum also runs tours of houses O'Keeffe owned.

792. *The Banjo Lesson*: American original. Henry Ossawa Turner (1859-1937) grew up in Pittsburgh. He eventually settled in Paris, where the African-American artist felt more accepted. His most recognizable painting, *The Banjo Lesson*, achieved fame because its depiction of an elderly man teaching a boy to play the banjo celebrated everyday life of African Americans, something ignored by other artists at the turn of the century. Go see it and a fine collection of African-American art at the Hampton University Museum in Hampton, Virginia.

791. Cheap Flights Abroad: *For the times when America is second best.* Yes, there are 1,000 reasons to love America, but sometimes people want to get away. A week in Paris is just the ticket, or maybe Mexico City, Bangkok, or London, which is often accessible on flights costing $99 one way. Too bad for people in far-flung locales, the inexpensive fares are usually from hubs like New York, Chicago, Miami, or Houston.

790. The Albuquerque International Balloon Fiesta: *Up, up, and away.* Launched in 1972, the largest hot air balloon festival in the world takes flight for nine days each October, filling New Mexico's skies with more than 500 floating orbs of all colors and designs.

789. MADD: *A sobering story.* Founded in 1980 by Candy Lightner, whose daughter Cari was killed by a repeat drunk driving offender, and Cindy Lamb, whose daughter Laura became the nation's youngest quadriplegic because of an inebriated driver, this nonprofit organization, Mothers Against Drunk Driving, has helped save thousands of lives by influencing the way we view this reckless behavior. With MADD's support, the national minimum age for purchasing alcohol was raised to 21 in 1984. The National Highway Traffic Safety Administration estimates this change has reduced fatal crashes involving drivers 18 to 20 years old by 13 percent and has saved an estimated 25,000 lives.

788. Fantasy Sports: *Leagues of their own.* Including this in the book was a close call. Time spent playing sports seems more profitable to the psyche than sitting in front of a monitor manipulating imaginary basketball, baseball, football, hockey, and NASCAR teams. Still, if you believe popularity is a measure of quality, consider this: 30 million Americans and Canadians take part in the pastime, the American Fantasy Sports Trade Association estimates.

787. John Deere Historical Site: *Home of the plow that turned the Plains.* In 1836 blacksmith John Deere revolutionized farming with a highly polished, contoured steel plow that cut clean furrows through sticky prairie soil. It stoked a manufacturing business that became the world's biggest producer of tractors. Today, visitors can tour the inventor's forge and homestead in Grand Detour, Illinois.

786. Angry Letters to the Editor: *A great way to get it off your chest.* For three centuries now American newspaper readers have capitalized on this tradition. If you've got a beef, put your tirade on paper, delete expletives (which turn off editors), and dash it off to your newspaper. It's your ticket to 15 minutes of fame.

785. St. John's College: *The last of the Great Books schools.* St. John's campuses in Annapolis and Santa Fe are throwbacks to a time when the curriculum in higher education centered on the great literary classics. As the school advertises on its Web site each summer: "The following teachers will return to St. John's next year: Homer, Euclid, Chaucer, Einstein, Du Bois, Virgil, Augustine . . ."

784. Honky-Tonk Bars: *Joe Six-pack heaven.* In these Saturday-night carnivals, the flag-lovin' crowd likes the beer ice cold and the jukeboxes loud. They are the stuff of detective novels, Westerns, and other thrillers. Take this scene from the opening page of James Crumley's *The Last Good Kiss*, published in 1978: "I slipped onto a stool between the bulldog and the only other two customers in the place, two out-of-work shade-tree mechanics who were discussing their lost unemployment checks, their latest DWI conviction and the probable location of a 1957 Chevy timing chain." Gimme another boilermaker.

783. Cape Cod Sunsets: *Picture postcards.* Some claim Key West, Florida, has the best sunsets. But we favor Cape Cod, Massachusetts, whose geography and location means the sun always sinks dramatically into ocean or bay, painting the horizon in spectacular pinks, purples, corals, and golds.

782. Hearst Castle: *Megalomaniac's interpretation of a "little something."* Nestled on 250,000 acres of prime California coastal real estate, the 165-room castle was built by publishing "king" William Randolph Hearst. He hired architect Julia Morgan in 1919, saying, " . . . we are tired of camping out in the open at the ranch in San Simeon and I would like to build a little something."

781. The Cody Firearms Museum: *Shooting gallery gone silent.* Love guns or hate them, they are in our blood and history, and the collection in Cody, Wyoming, ranging from Colt revolvers and

Winchester rifles to the M16 assault blaster, is as good a place as any to get a sense of their sweep and impact.

780. Kiawah Island: *Birdie haven, birders' delight.* With its numerous ponds, 40,000 acres of salt marsh and 16 miles of trails, this tropical isle near Charleston, South Carolina, also boasts five championship golf courses, including one with panoramic views of the Atlantic from every hole. Then there's the birding tours offering sightings of bald eagles and endangered wood storks.

779. Children's Board Games: *Bored children's games.* We concede that electronic toys have stolen the thunder from the children's toy market, but our littlest Americans still learn the joy of winning from their first game of Candy Land, how to lose gracefully by playing Sorry!, and how to count by playing Chutes and Ladders.

778. Saul Steinberg's Map of the World: *Prick to the pretensions of elitists.* Dominated by Manhattan's West Side, with the rest of the United States shoved into the background, and the Pacific Ocean, China, Russia, and Japan barely visible in the distance, the 1976 *New Yorker* cover was a hilarious jab at the provincialism of the city that likes to think of itself as the cultural capital of the world.

777. Fountain Favorites: *Memories in tall glasses.* Think back to root beer floats, the ice cream swimming in the soda, forming a fizzy head. What about egg creams, the chocolate syrup, milk, and seltzer combo with nary an egg in sight? You can still sip the former at A&W road stands and the latter, courtesy of Saul's restaurant in Berkeley, California, or a visit to Manhattan's Lower East Side.

776. Geeks: *Nerds with computer skills.* Thank goodness for these folks who can show us how to sort an Excel file alphabetically, reinstall a hard drive, or massage a software application. They may have been beaten up in high school, but geeks are the new cool guys.

775. Smith: *A name worth repeating.* If all 2.8 million people with the no. 1 surname in America lived in one state, it would be as populous as Nevada. Smith is the plain vanilla of names, but we bet most Smiths are happy not to be something zippier, like a Lipschitz or a Balasubramanium.

774. Dollywood: *Parton's playground.* Three million visit this Pigeon Forge, Tennessee, theme park annually, making it the biggest attraction in the state. Owned since 1986 by country music star Dolly Parton and the Herschend family, the park features 26 rides, a full-size steam engine, Dolly's tour bus, plus traditional crafts and music of the Smoky Mountains. The blond diva's also been known to sing there.

773. Sierra Club: *Protecting our planet, one issue at a time.* This grassroots organization founded in 1892 is our environmental conscience. Members live by their motto, "Explore, Enjoy and Protect the Planet." They campaign to encourage fuel economy and combat global warming. The club, with more than 700,000 members, also sponsors "outings," as they are called, to natural wonders to remind the adventurers that this wilderness they love is worth fighting for.

772. Hawaiian Shirts: *Perfect in Hawaii.* Except perhaps for hipster Jack Kerouac, men who wear those gaily patterned Hawaiian shirts have generally not been cool unless they were in Hawaii.

771. Silver Screen Sirens: *Ageless beauties.* From Greta Garbo, Jean Harlow, and Marlene Dietrich to Lana Turner, Elizabeth Taylor, and Marilyn Monroe, they dazzle with their incandescent sensuousness. Recent comers like Halle Berry, Cameron Diaz, and Angelina Jolie just up the ante. Men salivate over them. Women envy them. We all remember them—vividly. Just catch Rita Hayworth on YouTube in the joyously lusty, "Put the Blame on Mame, Boys." That's what we mean!

770. Book Clubs: *Giving the solitary delight of reading a group hug.* Book clubs, which bring people together periodically to discuss a specific work, have been around for at least a century. They surged in popularity in 1996 when talk show host Oprah Winfrey began hers. Media researchers described the "Oprah Effect," in which titles chosen for her monthly discussion became instant best sellers.

769. StubHub: *The heir to scalpers.* Founded in 2000, StubHub is the largest eBay-style online auction site for entertainment tickets. The company estimates the market for such resales is $10 billion a year. It's a safer way to get into a game or concert than buying tickets outside the arena from the guy with the shifty eyes.

768. Native American Crafts: *More than just baskets.* An antique handcrafted basket can be worth upward of $50,000 at auction, but Indian clothing, dolls, moccasins, dream catchers, rugs, blankets, ceremonial masks, weavings, chief's rattles, harpoons, headdresses, pipes, tomahawks, and beaded bags are all highly collectible.

767. The Great American Beer Festival: *We'll drink to that.* Since 1982, this annual Denver event has been the U.S. brewing industry's top public-tasting opportunity and competition. Nearly 50,000 quaffers can sample any of 2,000 different handcrafted products

from all over the country. Designated drivers get discount admission and soft drinks galore.

766. White Water Rafting: *Whoa mama!* Daring souls take canoes, kayaks, or inflatable rafts through rapids ranging from Level I (piece of cake) to VI (pray hard!). Somehow, thousands manage to survive and exhilarate in a breathtaking rush. One of the wildest rivers: the Chattooga, between South Carolina and Georgia, with Class IV+ rapids. This was a location for the movie *Deliverance*. Uh-oh!

765. The Adirondacks Park: *And you thought Yellowstone was huge.* This patchwork of public and private lands in upstate New York encompasses 6 million acres, almost twice the area of Death Valley National Park, the largest national park in the lower 48 states at 3.4 million acres (Yellowstone has 2.2 million). The size of Vermont, it has 85 percent of the wilderness in the eastern U.S., and with its mountains and misty lakes, is a paradise for tourists in summer and a hard slog for its 130,000 permanent residents in winter.

764. Old King's Highway on Cape Cod: *Perfect setting for a lazy afternoon drive.* Now called Route 6A, it began as an Indian trail and today winds for 34 miles past sun-glittering harbors, salt marshes and tidal flats, lakes and cranberry bogs, wooded parklands, sea captains' homes, lighthouses, windmills, church cemeteries, antiques shops, and 19th century inns. Drive it on a crisp fall day to catch the New England foliage, stopping as the fancy seizes you at any one of the restaurants, taverns, and historic sites.

763. The Corn Palace of Mitchell, South Dakota: *A-maizing place!* Inside, it's a 3,250-seat arena for concerts and conventions. Outside, the murals made of grass, grain, corn, and more corn are redesigned with a new theme every September to pay homage to South Dakota crops.

762. Outlandish Hoaxes: *Reassurance that some people are more gullible than we are.* The unearthing of the "Cardiff Giant," a 10-foot-tall petrified man behind a barn in 1869, a 1985 article in *Sports Illustrated* about a baseball player named Sidd Finch who threw baseballs at 168 miles an hour, and a 1995 film depicting the autopsy of a space alien, are classics. But the gold standard has to be the radio broadcast by Orson Wells in 1938 of a Martian invasion of earth that sent millions of Americans fleeing their homes and led to riots, looting, and clogged highways.

761. Loons: *Diving dynamos.* The common loon is the state bird of Minnesota, where 12,000 of these creatures thrive—more than in any other state except Alaska. Loons are built like torpedoes and can dive 250 feet underwater for prey. A bit smaller than geese, loons have long black bills and red eyes. In flight, they can clock 75 mph.

760. U.S. National Spelling Bee: *Spell-check be damned!* The advent of computer spell-check has pretty much put the kibosh on spelling as a marketable skill, but we still love watching this annually televised showcase that weeds out the best middle-school spellers from a pool of 11 million students. Frank Neuhauser of Louisville, Kentucky, won the first bee in 1925 by correctly spelling gladiolus, while the 2009 winner, Kavya Shivashankar of Olethe, Kansas, spelled *laodicean.* Try using that in a sentence. Hey, we just did!

759. A T-Bone Steak on a Grill: *To die for.* Eighteen ounces, rare and juicy, dotted with hot, crispy fat, with caramelized onions on top and a baked potato on the side. The favorite last meal of Americans facing the executioner.

758. Seventh Day Baptists: *Minuscule is beautiful.* They number just 5,000 in the U.S., a fraction of the Adventists, and a mere 50

in England, their birthplace. Fervent in their adherence to the Fourth Commandment to worship on the Sabbath, they are a reminder of the incredible diversity of religious affiliation in the United States.

757. Tipping: *Extra help for the help.* The word "tip" is an acronym for "to insure promptness," and unlike much of the rest of the world, where service fees are included in hotel and restaurant charges, we get to reward those underpaid waitresses, bellboys, and other providers for their efficiency and courtesy. The generous among us tip 20 percent of the bill. Of course, we reserve the right to stiff the hired help for rudeness, though we'd never sink as low as to drop a nickel in a glass of water to make our point, as some disgruntled diners do.

756. Frozen Dead Guy Days Festival: *Our wackiest celebration.* This annual event occurs in Nederland, Colorado, where a Norwegian man (who has since been deported) transported his grandfather's frozen corpse to a cryonics facility he and his mother planned to build there. That facility never got off the ground, so the frozen body is stored in a local shed. The festival includes coffin races, a polar plunge, and Frozen Dead Guy ice cream. Weird.

755. *Car Talk*: *Auto repair advice as entertainment.* In 1987, brothers Tom and Ray Magliozzi went national on their hilarious public radio call-in show for people with car troubles. It's funnier than most stand-up comedy, and you can learn the value of a well-maintained drive shaft flex joint.

754. Weird Research Results: *Conversation starters at dinner parties.* Have you heard about the study by the head of the Women's Studies Department at Radcliffe College that blamed country singer Patsy Cline's music for suicidal thoughts in some women? Your table companion wants to hear more.

753. **The Humana Festival of New American Plays**: *First footlights.* The Actors Theatre of Louisville, Kentucky, presents this annual event to thousands of theater lovers from 30 countries to preview the latest American offerings. More than 350 plays have been produced there since 1976, including Pulitzer Prize winners *The Gin Game, Crimes of the Heart,* and *Dinner With Friends.* Moreover, 90 million Americans have seen additional productions on stage or screen.

752. **Juilliard School**: *The easier way to get to Carnegie Hall.* America's no. 1 school for classical musicians and dancers—and drama students like Christopher Reeve, Robin Williams, and William Hurt—was founded in New York in 1905 so promising performers wouldn't have to travel to Europe to become virtuosi. Itzhak Perelman, Renee Fleming, and Yo-Yo Ma studied there, and so did jazzmen Wynton Marsalis and Chick Corea.

751. **Funny Epitaphs**: *Hidden cemetery treasures.* We like the one in Nevada that reads: HERE LIES A MAN NAMED ZEKE. SECOND FASTEST DRAW IN CRIPPLE CREEK. Then there was this, on a tombstone in Georgia: I TOLD YOU I WAS SICK.

750. **Teddy Bears**: *Plush comforters.* Our most popular stuffed toys, these cuddly playthings were introduced in 1902 and named for President Teddy Roosevelt after he refused to shoot a black bear cub on a hunting foray. Children like them for snuggling, and hospitals favor them for their bedside manner in calming young patients.

749. **Bisquick**: *Magical mix.* Introduced by General Mills in 1931, Bisquick provides premixed flour and fat for countless recipes. It can be transformed into just about anything—cakes, dumplings, pancakes, casseroles, muffins. We like the Savory Roasted Vegetable Strata and the Cherry Chocolate Chip Scones, two recipes that earn five spoons on the Bisquick Web site!

748. Iditarod: *Mush ado about something.* It started in 1973: a dog sled race across mountains, tundra, forests, and frozen rivers, 1,150 miles from Anchorage to Nome. The trek honors Alaska's frontier tradition and the memory of Balto, a dog who led a vaccine-carrying expedition that saved Nome from a diphtheria epidemic in 1925. Bronson led the speediest run, eight days and 22 hours in 2002, with Martin Buser at the reins. Iditarod means "a far distant place" in the Athabaskan Native American language and is the name of a river near the course. The winner each year gets $69,000 and a new truck. Organizers say care is taken to protect the dogs, but People for the Ethical Treatment of Animals would like to put the event on ice.

747. Crowd Control: *We do it better.* Yes, a Wal-Mart employee was trampled to death by shoppers on Black Friday 2008. And yes, 21 people died in a stampede in a Chicago nightclub in 2003. But the U.S. is generally spared the disasters that bedevil sports stadiums and religious celebrations in many countries, like the 1,400 Muslims crushed to death in a tunnel in Mecca in 1990. Check out Times Square on New Year's Eve (from the safety of your living room) and marvel at how 1 million plus delirious celebrants squeezed into a few blocks can all saunter away from the party unbruised. Thanks, NYPD; you set the standard for the world.

746. Ozark Mountain Country: *Old grounds, new sounds.* The backwoods of southern Missouri and northern Arkansas have mountains, rivers, forests, farms, and remnants of villages reminiscent of Dogpatch. Tourists tired of all that scenery can stop in Branson, Missouri, one of the busiest music hot spots in the world.

745. Hurricane Hunters: *In the belly of the beast.* June through November, hurricanes threaten the East and Gulf Coasts of the United States. For over 50 years National Oceanic and Atmospheric Administration experts as well as Air Force reservists have flown reconnaissance aircraft into these maelstroms, and their meteorological measurements give landlubbers a much appreciated heads-up.

744. *Green Eggs and Ham* and Other Oddities: *The books of Dr. Seuss.* He gave us Yertle the Turtle, Marvin K. Mooney, Horton the elephant, the Grinch, and the Cat in the Hat. Theodor Seuss Geisel (1904-1991), who wrote under the sobriquet Dr. Seuss, authored more than 60 children's books and demonstrated the depth of his imagination through odd-looking illustrations and quirky rhyme schemes. Geisel was not above creating words, like "Grinch" to describe a grouchy person, or "sneedle" and "ga-fluppted." His most famous book, *Green Eggs and Ham*, was written on a bet with editor Bennett Cerf, who challenged him to write a book using only 50 words. He did, and Sam-I-Am lives on.

743. The Chicago Cubs: *Futility with style.* Whether the Cubs win or lose—mostly lose—it's always "a bee-yooo-tiful day for baseball" at Wrigley Field, as announcer Harry Caray used to say. The Cubs play at home in daylight most of the time in an ivy-walled stadium built in 1914, and the fans are the most loyal in the sport, the most knowledgeable, and perhaps the most beer besotted. They root for the team with the most famous losing streak in baseball: no World Series trophy since 1908, when Tinkers, Evers, and Chance were turning double plays.

742. Instructions for Dumb People: *Timely reminders of our need for legal reforms.* Only in America does even the remotest threat to consumers' health and safety have to be declared. Everybody's heard about the origin of the "Caution: coffee may be hot" warning, but how about this one: "Beware: sled may develop high speed under certain snow conditions." For sheer zaniness, we nominate the guidance printed on an airline's packet of nuts: "Open packet, eat nuts."

741. Jackson, Wyoming: *Gateway to grandeur.* Millions hit Jackson en route to Grand Teton and Yellowstone national parks. A town of 8,600 full-time residents, it also is a major skiing destination. Jackson Hole Ski Resort, for example, boasts a slope of 4,139 feet, one of the highest vertical drops on the continent. If that doesn't grab you, try walking through the town's entrances: large arches made of elk antlers.

740. American Dogwood: *A gift of springtime.* After the first blush of forsythia and azalea, the more subtle dogwood graces our yards. Its pink or white flowers are actually colored leaves. We welcome their presence on suburban lawns everywhere spring springs up.

739. Hyde Park: *Incubator of the New Deal.* Franklin Roosevelt lived here on and off all his life, and seeing his rooms, furniture, and memorabilia is to understand how a 19th century patrician became America's great 20th century president. Visitors learn what a mama's boy he was. His mother Sara had the bedroom next door, and his wife Eleanor's was down the hall.

738. One Liners: *National horse laughs.* Here's a sampling: "All those who believe in psychokinesis raise my hand" (Steven Wright). "The more you read and observe about this politics thing, you got to admit that each party is worse than the other" (Will Rogers). Also, "100,000 sperm and you were the fastest?" (T-shirt slogan). And, "I refuse to join any club that would have me as a member" (Groucho Marx).

737. The Boeing 737: *Workhorse of the heavens.* In continuous production since 1968, the 737 is the most popular aircraft of all time. Boeing has won more than 6,000 orders for the cash cow's nine models (the 6,000th was delivered to the International Lease Finance Corporation on April 16, 2009), and today, more than 1,200 of them are plying the sky at any given time.

736. Tall Ships Parades: *Celebrating our maritime heritage.* They floated majestically up the Hudson River during the Bicentennial in 1976. Ever since, regattas of American tall-masted sailing vessels and international counterparts have drawn huge crowds to ports like Norfolk, Newport, Tacoma, and Boston for annual waterfront fetes. San Francisco's Festival of Sail has included USCG *Eagle*, a spectacular, three-mast Coast Guard training ship, and HMS *Bounty*, built for the film *Mutiny on the Bounty*.

735. Barn Owl: *A loner that doesn't give a hoot.* Owls are not smarter than the average bird, despite the "Wise as an Owl" simile. Nocturnal birds of prey, they tend to keep to themselves (although a group of owls is known  as a "parliament"). The barn owl, so named because that's where it likes to live, is the most common owl in the United States. It doesn't hoot, it screeches. It can be counted on to rid its habitat of rodents—worth keeping one in the barn for.

734. Boulder: *Brains and bikes.* Among the 180,000 adult residents, 30 percent have undergraduate degrees and five percent have doctorates. That makes the home of the University of Colorado the most educated city in the United States, according to a *Forbes* magazine study. Other surveys put Boulder in the top 25 among bicycle-friendly cities, healthy places to retire, and towns near high-tech jobs. For plain folks who want a break from their liberal, elitist neighbors, the Rocky Mountains are in Boulder's backyard and Denver is 30 miles south.

733. The Sylvia Plath Industry: *Tabloid heaven for literary snobs.* Plath, a minor talent, casts a far greater spell on Ivy League intellectuals than Robert Frost, America's favorite poet. And why not? Her tempestuous marriage with British poet Ted Hughes that ended in her suicide in 1963 created an ever-evolving retrospective of passion, betrayal, and repentance that has been mulled over by one biography after another. "Them again," lamented a *New York Times* review of one such book in 2003. "Just when you thought there was no more to be said, the ransacked remains of Ted Hughes and Sylvia Plath float to the surface once more."

732. Chris-Craft: *Great way to float your boat.* Recreational cruising scores high with Americans, probably because 90 percent of us live under an hour from some navigable body of water. Some 70 million Americans used onboard pleasure craft in summer 2008, five percent more than in 2007. For its craftsmanship and engineering, Sarasota, Florida's Chris-Craft claims the title of our premium brand.

731. Phi Beta Kappa: *Key to academic success.* This most prestigious academic honor society was born the same year as the United States. About a half million members strong, ΦBK counts 17 former presidents and 131 Nobel Laureates as members. ΦBK is the ultimate in exclusivity among the crème de la crème: you have to be invited to join.

730. Graceland: *The building Elvis left.* Elvis Presley bought this Memphis mansion in 1957 from a tycoon who had named it for his daughter. Now it's a museum, where hundreds of thousands of tourists a year learn about Elvis's life, see his gear, and visit some of

the rooms. Bad news for the ghoulish: the second floor bathroom where the drugged-up king of rock 'n' roll died is so far off limits that many of the employees have never been there.

729. UFO Sightings and Alien Abductions:
Symptoms of our intergalactic wanderlust.
These aren't hoaxes; they are stories told by true believers. The most famous UFO incident was the crash landing of a space vehicle in Roswell, New Mexico, in 1947, but millions more accounts have been recorded. After all, 13 per-

cent of Americans say they have seen flying saucers, and 32 percent believe the elusive ships are real. As for the wave of abductions by extraterrestrials in recent years, the victims' harrowing accounts have been so detailed that John Mack, a Harvard University psychiatrist, called the encounters an "authentic mystery."

728. *The Sound and the Fury*: *Brooding tale of the post–Civil War South.* Often cited as Nobel prize winner William Faulkner's finest work, this 1929 novel uses the main characters' internal monologues—sometimes disordered or insane—to chronicle the decline of the Compsons, an aristocratic family of fictional Yoknapatawpha County, Mississippi. Faulkner's style isn't easy, with his characteristic run-on sentences, frequent flashbacks, and stream-of-consciousness, but he delivers a powerful emotional punch.

727. Free Stuff: *All the swag you can bag.* From ballpark bobble heads to goodies at birthday bashes, getting something for nothing is a great delight even when the gift is a soggy grocery store meatball on a toothpick. The pinnacle of freebie frenzy is the bestowing of luxury items to Oscar nominees and presenters. Things like a year's supply of Manni olive oil and weeklong stays in Hawaiian resorts. The best free stuff is what we get serendipitously. Like when we're driving down a road and spot furniture, wood, old books, or a beat-up kayak in a front yard with a sign: FREE.

726. Ranches: *Endangered heritage.* A generational gap exists on U.S. ranchlands, where graying old-timers far outnumber young cowpokes. With high land prices, poor owner-succession strategies, and the lure of cities, agricultural states are losing ranchers and cattle, posing a threat to the wide-open spaces and jeopardizing our food supply. In 2008 the number of beef cows dropped to 41.8 million, an all-time low. The number of ranches and other types of beef-cow operations has decreased to fewer than 758,000, a 22 percent decline over the past two decades. Still, the tide could turn with ideas like Nebraska's 100-cow program, which helps hopefuls get low-interest government loans once they complete a ranch management course.

725. Taxpayer Revolts: *Take that, IRS!* You could argue that this country was founded on the basis of a tax revolt. It all started when angry colonists heaved three shiploads of tea into Boston Harbor in 1773 to protest England's Tea Act as taxation without representation. Protesting against high taxes has been an American political pastime ever since, with notable modern-day episodes including California's 1978 voter-approved Proposition 13 that capped property taxes—and contributed to that state's bankruptcy—Ronald Reagan's antitax ascendancy, and the recent backlash against ballooning national debt known as the Tea Party Express.

724. Hip Hop: *Rapping with style.* Hip hop sounds best blaring through a boom box. Music historians trace the history of modern Hip Hop to the rhymes of Muhammad Ali. Performers like Ice T, the Beastie Boys, Dr. Dre, and Notorious B.I.G. proved to us that Hip Hop is more than just the music, it's a lifestyle.

723. The Arkansas River: *A slow flow east.* The 1,450 mile river starts in Colorado and ends in Arkansas, and the best part is the newly navigable section between the Mississippi River and Tulsa, home of the largest ice-free port on the entire 25,000 miles of inland U.S. waterways.

722. Weird Inventions: *Fantasy ticket to riches*. Whether a golf club that is propelled by the release of high-pressurized water, a fork that flashes a red light to tell you to stop eating, a voodoo-doll toothpick holder, or something as simple as an indestructible Christmas-tree stand as heavy as a cinder block, the cockeyed brainstorms of amateur innovators never fail to surprise. Some of them do make their progenitors rich; inventor Ronal Popeil became a multimillionare pitching products like the Inside the Egg Scrambler on TV. With the number of patent applications by Americans doubling in 10 years, to 240,000 in 2007, more fun is on the way.

721. Salt Water Taffy: *A chew for all seasons*. Originally, this corn syrup and sugar confection was made and sold seaside in summertime beginning in 1883. Likely our first mass-produced candy, today it is marketed year-round, coast to coast, in flavors as toothsome as eggnog and guava.

720. Keck Observatory: *Star Wares*. What are eight stories high, weigh 300 tons, and track the galaxies, sometimes with the clarity of the Hubbell space-based star gazer? The twin telescopes of the Keck Observatory in Hawaii, the largest optical and infrared machines of their kind in the world. They captured the first images of planets around another sun and have been mapping a large black hole in the Milky Way. It's ironic that a great observatory pointed to the skies was funded by the estate of an oil tycoon who invented ways to burrow into the earth.

719. *Star Wars*: *Special effects dazzler*. Consistently cited as one of the best U.S. films ever made, George Lucas's 1977 space adventure used groundbreaking visual pyrotechnics to transport us to a distant galaxy where freedom fighters battled the Death Star space station of the oppressive Galactic Empire. In this, and five sequels, we met indelible characters Luke Skywalker, Darth Vader, Princess Leia, and Jedi Master Obi-Wan Kenobi. A record-maker at the box office—the film earned $460 million here and $337 million overseas—it influenced dozens of future filmmakers and convinced Hollywood that fast-paced, big-budget blockbusters were the future.

718. Tie Dye: *Hippie haute couture.* This could have gone the way of headbands, peace symbols, and fringed vests. Tie-dye clothing, especially T-shirts, has grown from an at-home craft project into high fashion. It's easy and it's fun. A box of dye, some rubber bands, a white T-shirt, and easy-as-pie Internet instructions are all you need to make your own. Don't forget to wear rubber gloves!

717. Cheerios: *The unsoggy.* What General Mills calls the first ready-to-eat oat cereal started as Cheeri-Oats in 1941. We like 'em because they don't get mushy in a bowl of milk as quickly as their peers, Rice Krispies and Corn Flakes.

716. Haunted Dwellings: *A growth industry in times good and bad.* Almost half of Americans believe in ghosts, and one in five has seen them. Aside from houses, the United States boasts haunted hotels, nightclubs, lighthouses, fields, prisons, museums, hospitals, battlefields, and even entire towns, like Athens, Ohio. Entrepreneurs have cashed in on our fascination with thuds in the night, pianos playing by themselves, flickering lights, and bleeding apparitions by organizing guided tours of the infested dwellings. In sum, poltergeists are no match for Yankee capitalists on the lookout for a quick buck.

715. Hairpiece Charities: *Covering losses.* Children who suffer permanent hair loss due to medical conditions can receive free or low-cost natural hair wigs courtesy of nonprofit groups like Locks for Love and Wigs for Kids. More than 80 percent of the donated hair comes from children. All that's required are strands 10 inches or longer—a good-sized ponytail. As sponsors say, the reward's in knowing that when kids feel better about their looks, they feel brighter about their futures.

714. The Property Rights Movement: David v. Goliath. A 2005 U.S. Supreme Court ruling strengthened government authority to condemn private property for development by private, for-profit investors. The decision alarmed owners already wary of eminent domain—the legal process that allows governments to take privately held land for uses deemed to benefit the public. Critics such as the Institute for Justice, the libertarian public-interest law firm, say this practice is fraught with abuse. Their grassroots tactics have helped citizens trounce questionable land grabs across the nation and encourage 34 states to limit the ability of local and state governments to seize private property.

713. Fulbright Programs: *Our brain import-export business.* Since 1946 this program, originated by the late Arkansas senator William Fulbright, sends U.S. students and scholars abroad and brings their foreign counterparts here. In 2007, Congress appropriated $262 million to fund about 6,000 exchanges to further cultural understanding.

712. Taking a Number: *Peace at the counter.* The numbers on the little tickets dictate who is next. No lines, no scrums, no shouts of "Hey, I was before you."

711. The Trust for Public Land: *Tree-huggers' champion.* This national nonprofit works with communities and environmentalists in landscapes from wildernesses to inner cities, to conserve land for parks, gardens, and forests. Since its founding in 1972 the trust has saved more than 2.5 million acres from bulldozers. Top priority: making cities greener.

710. Unusual Town Names: *Eccentricity, American-style.* We like Ding Dong, Texas; Possum Grape, Arkansas; Lizard Lick, North Carolina; Dunmovin, California; Yreka Zzyzx, California; No Name, Colorado; Zap, North Dakota; Buck Snort, Tennessee; Gnaw Bone, Indiana; and Climax, the name of municipalities in Georgia, Michigan, Minnesota, New York, North Carolina, Ohio, and Pennsylvania. Oh, we almost forgot, there's also Intercourse, Pennsylvania.

709. Pick-Your-Own Farms: *C'mon down and harvest.* Enterprising growers have realized it can be cheaper to let townsfolk traipse through the fields to bring home corn, berries, and pumpkins than to hire migrant labor. But the public wins big too: there's nothing like the aroma of fresh-picked apples on a crisp autumn afternoon.

708. The Battle for Clean Air: *One breath at a time.* The air we take in has improved since 2000, attests the Environmental Protection Agency. Thanks to the Clean Air Act, approved in 1970 and amended in 1990, we have a framework for national, state, and local efforts to monitor air quality and enforce standards. We shine in reducing ozone levels and resultant smog in our cities. Unleaded gasoline has helped cut lead particulates in the atmosphere. The downside is we still rank as the free world's biggest carbon producer. But we're working on that, even as emerging giants China and India churn out more of the stuff by the day.

707. Glacier National Park: *Grizzly country.* Its 1,600 square miles in northern Montana make it bigger than Rhode Island, and 700 miles of trails make it a hiker's paradise. The camping season is short because of snow, but that will change. By 2030, global warming will melt all the glaciers in Glacier National Park, the National Park Service warns.

706. Lost-Cause Crusades: *Grist for journalists on slow news days.* Movements to make English our official language, eliminate the penny from circulation, and remove fluoridation from our drinking water are doomed to failure, but they will never die. They won't even fade away.

705. USA Softball Women's National Team: *Bringing home the gold.* The American women have dominated the International Softball Federation since its founding in 1965. The team has six international gold medals for the world championships and has won two of the three Olympic gold medals ever awarded for softball. Alas, the sport has been eliminated from the Olympic roster for 2012. Meanwhile, American bats keep swinging successfully around the world.

704. Corvettes: *Chevvies with zip.* Since the first one rolled off a GM production line in 1953, the Corvette has been America's premier sports car—a vehicle for guys who want to attract chicks. Rich guys. In recent years the top model started at $104,000.

703. The Bulwer-Lytton Fiction Contest: *Consolation for struggling writers.* The competition for the worst possible opening lines for a novel, sponsored by the English Department at San Jose State University and named after a 19th British baron who opened one novel with the infamous phrase, "It was a dark and stormy night," makes anything you have written seem the height of elegance. Winning the prize—you write just a paragraph, not a whole book—takes imagination. A sample howler: "Theirs was a New York love, a checkered taxi ride burning rubber, and like the city their passion was open 24/7, steam rising from their bodies like slick streets exhaling warm, moist, white breath through manhole covers stamped 'Forged by Delaney Bros., Piscataway, N.J.'"

702. The Westminster Kennel Club Dog Show: *Where winners woof it.* Established in 1877, Westminster is the oldest U.S. club dedicated to the sport of purebred dogs. Canine fanciers drool over its annual all-breed show televised from New York's Madison Square Garden, where nearly 2,500 animals compete. Best in Show 2009: a Sussex spaniel named "Ch Clussexx Three D Grinchy Glee," a.k.a. "Stump."

701. The Yo-Yo: *The whirl on a string.* This toy goes back centuries, but modern yo-yoing really took off in the 1920s when Donald Duncan bought a popular yo-yo shop in Santa Barbara, California, and marketed two weighted pieces of wood, connected by an axle and powered by a long string. Today, international competitions feature high-end models that shatter world records. One example: a "sleeping" yo-yo, lolling at the bottom of the string but still spinning, has been twirled for 16 minutes.

700. Scary Movies: *Something to scream about.* Ever since film pioneer and inventor Thomas Edison made an 18-second movie, *The Execution of Mary Stuart*, in 1895, Americans have relished being scared to death in movie theaters. Our nominees for the scariest ones ever include *Invasion of the Body Snatchers* (the 1978 version), *The Shining*, *The Texas Chainsaw Massacre*, *Psycho*, *Silence of the Lambs*, and *The Thing*. Warning: don't watch any of these when you are home alone.

699. Cadillac: *The Cadillac of American cars.* The first was built in 1902, and the brand has survived depressions, recessions, and $4-a-gallon gasoline because Cadillacs are like the country: big, powerful, and immodest. The President got a new one in 2009 that weighs almost eight tons and has eight-inch-thick armor plate.

698. The Mormon Tabernacle: *Early megachurch.* Completed in 1867 to seat 8,000 during meetings of the Mormon Church, this building in Temple Square, Salt Lake City, Utah, won praise from noted architect Frank Lloyd Wright as "one of the architectural masterpieces of the country and perhaps the world." Still used for overflow crowds during the church's General Conference, the house of worship is also home to the world-renowned, 360-voice-strong Mormon Tabernacle Choir.

697. Beach Books: *Everything under the sun.* Don't make us read *Crime and Punishment* or some obtuse Faulkner tome on a beach with a million distractions. Give us a juicy novel that will help us forget just how hot and uncomfortable it is on the sand. We especially like authors John Grisham, Tom Clancy, Pat Conroy, Harlan Coben, Mary Higgins Clark, Lisa Scottoline, Ken Follett, and James Patterson.

696. Hard-Boiled Detectives: *The ultimate fantasy of Walter Mittys everywhere.* Born in the pulps of the 1920s and 1930s, these tough-talking ("Dust, pal, or I pump lead"), hard-drinking, quick-thinking, wisecracking loners who roam the mean streets to mete out rough justice and fall under the spell of femme fatales found their highest literary expression in Dashiell Hammett's Sam Spade and Raymond Chandler's Philip Marlowe, though their modern-day heirs, like Patricia Cornwell's medical examiner Dr. Kay Scarpetta, do pretty well for themselves.

695. The USS *Arizona* Memorial: *Shrine to our Pearl Harbor dead.* The devastating Japanese air attack on the U.S. Pacific fleet in this Hawaiian port on December 7, 1941, propelled us into World War II. Dedicated in 1962, this 184-foot-long memorial spans the mid portion of doomed battleship *Arizona,* now a watery tomb for many of its 1,177 crew members. As the memorial's designer observed:

it sags in the center but stands strong and vigorous at the ends, expressing "our initial defeat, then ultimate victory."

694. *Oprah*: Still *talking*.
Oprah Winfrey agreed to continue her popular hour-long afternoon talk show until September 2011. The highest rated talk show ever, when it debuted in 1986, *Oprah* went head-to-head with *The Phil Donohue Show*. He bowed out in 1996, but she kept on talking. Her mostly female viewers empathize with the charismatic Oprah, who has shared her own seesaw weight battle and the sexual assault that paralyzed her as a child. Oprah gets the guests she wants: celebrities like Tom Cruise and John Travolta; President Obama; people in the news. An interview with Oprah can make you an instant celebrity. Or, if she chooses your book for her TV book club, it's guaranteed best-seller status. When Oprah talks, people listen.

693. Verbal Treasures: *Pithy pronouncements that define us.*
First, money: "A penny saved is a penny earned" (Benjamin Franklin); "The business of America is business" (Calvin Coolidge); "What this country needs is a good five-cent cigar" (Vice President Thomas Riley Marshall); and, especially relevant in today's consumer culture, "There's a sucker born every minute" (Attributed to P. T. Barnum). Second, politics: "Mr. Gorbachev, tear down this wall" (Ronald Reagan). Finally, the sense of the absurd: "It was déjà vu all over again" (Yogi Berra).

692. Skyline Drive: *105 miles of almost heaven.*
Skyline Drive runs the north-south length of the Shenandoah National Park through the Blue Ridge Mountains of Virginia. It's breathtaking year-round, but October's foliage is so spectacular it can create gridlock alert on the 35-mile-per-hour, stop and smell the flowers highway on weekends.

691. The Spectacle of Politicians Self-Destructing: *Watching winners turn sinners.*
New York governor Eliot Spitzer shamed

himself by consorting with a hooker and resigned in 2008. Governor Mark Sanford of South Carolina jetted to a South American sex tryst in 2009 and before long was facing impeachment proceedings. But 2008 Democratic presidential candidate John Edwards wins the smarm award. Revelations the following year that he fathered a love child with a campaign worker behind his cancer-stricken wife's back ended his political career. Every time a big-shot lawmaker gets caught with a loose zipper, the masses are outraged for a couple of minutes. Then they switch the channel to another reality show.

690. How-To Books: *Teaching tomes.* Maybe it's their effervescent can-do spirit, but Americans relish the idea of improving their skills through books, be they gourmet cooking guides or woodworking manuals. Interests mirror the times. In the white-hot housing market, people clamored for titles about buying property and making money in real estate. But by early 2009's deepening recession, publishers were promoting advice books for consumers burned in the housing game.

689. Hull House: *Detour off the beaten path.* People flock to Chicago to visit the country's tallest building, frolic in Lake Michigan, root for the Cubbies, and ride the El. We propose you veer from the ordinary and take the no. 7, 8, or 60 bus from downtown to learn about Jane Addams and her ministrations to Chicago's poor. In 1889, Addams founded this "settlement house," as it was known, to provide education, guidance, and physical assistance to Chicago's poor women and children. Nearly 120 years later the Hull House Association proves good people can make a difference by annually helping 60,000 needy Chicagoans.

688. American Beauty Rose: *Haunting hybrid.* A rose by any other name just doesn't smell as sweet. Brought to the United States from France in 1875, this hard-to-find rose boasts heavily perfumed, crimson, voluptuous blooms. It inspired a Frank Sinatra song, appeared often in the 1999 film *American Beauty*, and graced a Grateful Dead album cover.

687. *Titanic* (the Movie and the Metaphor): The sinking ship.

OK, so it was built in Belfast and sailed out of Southampton, England, on its maiden voyage in April 1912 when it hit an iceberg and went down. We still think it has a place among America's best things. It has endured nearly a century as an overused metaphor of a lost cause. Most especially, Hollywood produced the 1997 film epic, *Titanic*. At 194 minutes, the movie almost lasted longer than the ship's only voyage. The ship cost $7.5 million to build, while the film cost $200 million to make. Still, the ship was a real loser, but the movie won 11 Academy Awards, including Best Picture, and grossed $1.8 billion worldwide.

686. Gadflies: Shooters from the lip.

Socrates saw himself as an irritating, biting gadfly, doggedly challenging authority, but "all in the service of truth." Gadflies can bug us, but their letters-to-the-editor and relentless questions at municipal and shareholders' meetings revitalize democracy. Where would we be without consumer watchdog Ralph Nader or pushy bloggers?

685. South Beach: Miami's funky spot.

Any place that is warm in January is appealing, but Miami's South Beach earns special mention for hip nightlife, Art Deco architecture, couture boutiques, and pristine beaches. By day, it's where you go to shop and sun-

bathe. By night, it's where you go to party and be seen. Europeans and South Americans love it. We do too.

684. Texas Water Safari: Wet 'n' gritty.

The annual 260-mile paddle race down the alligator-and-snake-filled San Marcos and Guadalupe Rivers is as tough as it gets. The 100 or so contestants must carry everything they need to survive the journey except for ice and water, which are provided along the way. The record time of 29 hours was set in 1997, but just finishing within the 100-hour time limit is a major feat. And there is no prize money.

683. The Big Dig: *Expensive lesson in urban renewal.* The $22 billion, 15-year project to reroute Interstate 93 from a hulking and traffic-choked elevated highway through the heart of Boston to an underground tunnel is one of the most daring, if most expensive, infrastructure projects of our age. Though plagued by cost overruns and sometimes shoddy construction, by the time it was finished at the end of 2007, the Big Dig had replaced a concrete horror with a mile-long pedestrian haven of lawns, trees, paths, and fountains.

682. The Big Pig Dig: *Fossil hunting at Badlands National Park.* Visitors to this site in the 381-square-mile park can watch paleontologists recover bones from one of the richest fossil sites in America, where animals roamed 33 million years ago. The name comes from one of the 18 species found there, a large piglike mammal, Archaeotherium.

681. The Golden Fleece Awards: *A rebuke to plunderers of the people's purse.* Initiated in 1975 by Senator William Proxmire of Wisconsin to throw a spotlight on taxpayer-funded projects like research into the drinking habits of sunfish and an $84,000 study on why people fall in love, the honorific continues to be invoked to shame the purveyors of political pork. The most famous recent example was the $300 million earmark for a "Bridge to Nowhere" in Alaska, a boondoggle, supported by then-Governor Sarah Palin, that succumbed to a growing public backlash against such giveaways and itself went nowhere.

680. Summer Stock: *Seasonal scenes.* Vacationing urbanites wanted high-quality theater. Shuttered Broadway shows wanted

income. Combine those needs, and voilà, the Straw Hat Circuit of indoor New England theaters was born in the 1940s. Future luminaries like Henry Fonda and Geraldine Page honed their skills there. Today, summer theater flourishes in hundreds of U.S. venues where current stars like Anne Hathaway and *American Idol*'s Taylor Hicks still tread the floorboards. So kudos to the Cape Cod Playhouse, Dennis, Massachusetts, our oldest, continuously operating professional summer theater.

679. Harley-Davidsons: *Hogs of the road.* First made in 1903, Harleys are to motorcycles what middle linebackers are to football: big, powerful, and surprisingly fast. Like the 994-pound Ultra Classic Electra Glide, which one reviewer called a brash behemoth. To see 100 years of Harleys, visit the company museum in Milwaukee.

678. Pine Valley Golf Club: *The country's best course.* Both *Golf Magazine* and *Golf Digest* have consistently chosen the 620-acre tract in the New Jersey Pine Barrens near Philadelphia as no. 1. Golfers love its challenges and dangers—all that water, scrub, and sand, all those thickets and steep drops—as well as its miracle-performing caddies, its scenic beauty, and its simple but graceful clubhouse.

677. Catholic Schools: *Prayerful powerhouses.* Since the 1840s, immigrant children in urban areas have received solid educations at little or no cost through Catholic parochial schools, the equal or better of many public schools. Learning transpired in a religious atmosphere, discipline was strict, and uniforms mandatory. But this once vibrant system is contracting. More than 3,000 Catholic schools have closed nationwide since the 1960s. Hiring more lay teachers to supplant dwindling numbers of religious has pushed costs up, boosting tuition out of the reach of many. Still, Catholic schools remain the largest nongovernment provider of education in this country, with students performing well on standardized tests.

676. Pacific Coast Highway: *A 485-mile breathtaking jaunt that links San Francisco and L.A.* We recommend you forego speed in favor of this twisting, harrowing ride along the rocky West Coast. A designated driver needs to keep eyes on the road while passengers ooh and aah over the crashing waves and gargantuan rocks that edge the Pacific Ocean.

675. Congaree National Park: *Primeval southland.* *National Geographic* placed this bottomland forest 20 miles south of Columbia, South Carolina, among the best national parks for canoeing. Paddlers can glide along Cedar Creek through stands of Bald Cypress and Water Tupelo draped with Spanish moss, accompanied by a symphony of songbirds, owls, and woodpeckers.

674. Preppy Dress: *Snooty way to fake it till you make it.* The author of this entry was a Hoosier rube who almost got hooted out of an East Coast private school for the sin of wearing seriously uncool clothing. He quickly learned that button-down Brooks Brothers pastel shirts, Bass Weejuns penny loafers, light blue cashmere sweaters, navy blue blazers, and tan chino pants were the tickets to acceptance by the dominant WASP culture, a basic ladder-climbing tactic that works to this day.

673. The Columbia Restaurant: *Delicioso.* Founded in 1905 to feed cigar-rollers in Tampa's historic Ybor City, its 15 dining rooms serve Spanish-Cuban classics like meltingly delicious Cuban sandwiches and authentic Valencian paella, paired with a wine inventory of over 30,000 bottles. Still family run, this is Florida's oldest eatery and the world's largest Spanish restaurant.

672. Punk: *Bad boys of rock.* Angry yet melodic rants against "the man" identify this genre. Punk, performed by the first garage bands, was popularized in the old CBGB club in New York. The Ramones and Green Day be damned, we're partial to the ravings of One Short Fall.

671. Floating Down a River in an Inner Tube: *Sport for Huck Finns.* Drift slowly and silently down a river and think about nothing. That's how to spend a few summer hours. Make sure the water is lazy and shallow, not like the Poudre River as it flows through Fort Collins, where a tuber without a life vest drowned in 2008.

670. Allocation of Calendar Chunks to Every Conceivable Cause: *Marketing gone mad.* We all know about Mother's Day and Secretary's Week, but Johnny Appleseed Day? National Playground Safety Week? Popcorn Month? We're not knocking this oh-so-American custom; we applaud it. As the French would say, "*C'est très folklorique.*"

669. Special Operations Forces: *Military muscle.* Best known of these elite, highly trained combatants are the legendary Green Berets, active since 1952. One example: in 2008 in an Afghan mountain valley, a dozen Green Berets jumped from a helicopter to ice-covered terrain and fought seven hours against hundreds of enemy. Supported by some Afghan commandos, they drew rifle, grenade, and machine gun fire. Carrying four wounded, the Green Berets fought their way to a chopper and escaped. They'd killed 200 terrorists.

668. Bing Crosby Music: *How every man thinks he sounds in the shower.* Der Bingle, as he was known, crooned his way through the best selling song ever recorded. His smooth, 1942 version of Irving Berlin's "White Christmas" has sold more than 100 million copies worldwide.

667. Jack Daniel's Old No. 7: *A sip of the Smokies.* We were disappointed to learn that Jack Daniel's distillery isn't owned by bearded guys in dirty overalls but is just another brand, alongside Korbel champagne, Finlandia vodka, and Bolla wine, of conglomerate Brown-Forman. Even so, the Tennessee sour mash whiskey is among America's finest liquors. It tastes like Kentucky bourbon, only better.

666. Bengay: *A cream that rubs us the right way.* Ubiquitous in locker rooms, it eases joint and muscle pain in arthritic oldsters as well as athletes. Ahhh! We've all whiffed its powerful menthol odor since we were kids, though these days you can buy "vanishing scent" gels. If you don't like goo, it also comes in patch form.

665. The Wild Turkey: *Shoulda been a contender.* Ben Franklin thought this gobbler, not the eagle, deserved to be the national bird. Even now the eagle gets better press. But turkey, our largest game bird, was an important food to Native Americans.  These scarlet-necked fowl, scarce by the early 1900s, were successfully reintroduced after World War II. It's their domestic cousins—more than 45 million of them—that we carve at Thanksgiving.

664. Quilts: *Patchwork warmth.* In theory, a quilt is a bedcover made to keep you warm. But in reality, colorful American quilts, fashioned by women at quilting bees through the centuries, are works of art. Stitched into intricate designs using bits of fabric remnants, treasured antique quilts are usually hung, like artwork, rather than gracing a bed.

663. LPs: *What goes around comes around.* First, long-playing records were supplanted by audiotapes. Then came compact discs, followed by iPods. Someday we will listen to music on chips, planted in our brains at birth, containing the sound from every recording since Thomas A. Edison's wax cylinders. In the meantime, vinyl is coming back and LPs in the attic are cool again, though kinda scratchy. Like the one playing while this is being written: A 1968 RCA Victor record of pianist Vladimir Horowitz.

662. Paul Bunyan and Babe the Blue Ox: *Giants of folklore.* The Minnesota lumberjack was so huge and powerful he combed his hair with a pine tree and dug Lake Michigan as a drinking hole for his animal friend. The tall tales about these two companions of the northern forests, embodying the frontier virtues of macho bravado and can-do tenacity, have fascinated children (and tested the inventive powers of their parents) for 100 years.

661. Tiffany Glass: *Iridescent treasure.* Frustrated decorator Louis Comfort Tiffany opened his own foundry in 1878 to produce the quality glass he couldn't find elsewhere. The result: richly colored and textured handmade glass that, on his popular butterfly lamps, for example, resembles woody branches, flowing water, and sun-dappled foliage. Today, collectors hoard the extraordinarily expensive originals. Others buy replicas for a mere $300.

660. Collard Greens: *Eat your vitamins.* There are those who allege that anything cooked in pork fat tastes good. Collard greens are Exhibit A. Collard greens, a leafy member of the cabbage family, are loaded with Vitamins A, K, and C and provide a good source of calcium, folate, and manganese. They are a staple on holiday tables in the South, especially on New Year's Day, when they are eaten to ensure wealth in the coming year.

659. **Antique Car Displays**: *History on wheels.* At one end of the spectrum are collections like the 130-auto Gilmore Car Museum in Hickory Corners, Michigan, where visitors can have fantasies of being Gatsby or Edith Wharton out for a drive in a 1926 Wills St. Claire. Then there are the fund-raisers at the local firehouse, where a model-T Ford may stand alongside cars that just qualified for antique status at the age of 25. Like a 1985 Dodge Caravan restored to mint condition. These are the cars that move us emotionally, because we remember driving them.

658. **New York City Taxi Drivers**: *Attitude you can believe in.* They come in all ages, sizes, and accents. Did we mention opinions? A lot of visitors would be lost, literally, without them. Some natives hate them, but if the trademark yellow cabs vanished overnight, pickup trucks and SUVs from the suburbs would clog the Big Apple in their place.

657. **Cape Hatteras Light**: *Beacon of safety.* One of the few traditional U.S. lighthouses still in operation, this North Carolina tower stands 208 feet, making it the tallest on our coasts. For over 100 years its warning beams have saved many a ship from the treacherous Diamond Shoals extending 14 miles into the Atlantic. Painted in black and white spirals, the lighthouse has been dubbed "the Big Barber Pole."

656. **Cecil B. Day Butterfly Center**: *Making your heart flutter.* More than 1,000 tropical butterflies flit through the conservatory at Calloway Gardens in Pine Mountain, Georgia. The center is named for the founder of the Day's Inn motel chain. There are other insect and butterfly museums around the country, but this one is the monarch.

655. Snow Days: *Nature's gift to goof-offs.* The fantasy starts forming the night before when the flurries begin, and by morning the good news begins to spread by radio, telephone, or Internet. No school today. The office is closed. Roads closed to essential traffic only. God is good.

654. The Masterpieces of Edward S. Curtis: *Haunting images of a forgotten way of life.* Curtis took more than 40,000 photographs of Native Americans from dozens of tribes in the early 20th century, recording for posterity scenes ranging from dancing in a forest to gathering reeds in a fog-shrouded lake that stand today as a testament to disappearing cultures of the people who got here first.

653. Rock and Roll Hall of Fame and Museum: *The beat goes on.* Elvis Presley's jukebox. Jimi Hendrix's electric guitar. Janis Joplin's fringed scarf. These artifacts and more, plus films, music, and photographs, give us a close-up glimpse of some 200 celebrities, including artists, producers, and songwriters, whose work has been honored at this Cleveland site since 1986.

652. Clams on the Half Shell: *Slimy, yet satisfying.* Order a dozen Little Necks. Sprinkle with lemon juice, dab with cocktail sauce (a mixture of ketchup and horseradish), then slurp that sucker out of the shell and down the hatch.

651. *Casablanca*: *Art rising from confusion.* The actors in this 1942 Oscar winner hated director Michael Curtiz. The producer for Warner Bros. thought up the final, brilliant, "I think this is the beginning of a beautiful friendship," after shooting, and Humphrey Bogart dubbed it in. Also dubbed was the piano in "As Time Goes By." Dooley Wilson was a drummer in real life and he finger-synced a pianist playing off camera. Ingrid Bergman was spooked because throughout most of the shooting she didn't know the ending. Neither did the studio, according to some accounts. "Warner had 75 writers under contract and 75 of them tried to figure out an ending!" coauthor Julius Epstein told *Hollywood Hotline* 50 years later. The story about Nazis, World War II refugees, freedom fighters, love, and sacrifice may not amount to a hill of beans in this crazy world, but we love it.

650. Lake Mead: *Biggest manmade body of water in the U.S.*

Now and then, Lake Oahe in the Dakotas tries to grab that title because it covers more space, 600 square miles versus Lake Mead's 248, but Mead, created 30 miles south of Las Vegas by the Hoover Dam, contains more water, 28.5 million acre feet versus 23.5 million. Size matters—which is why we worry about predictions the

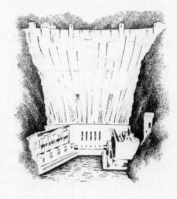

reservoir could dry up by 2021 if arid conditions persist and drastic measures aren't taken to reduce water use in the region.

649. Skateboarding: *Dude, it's the thrill!*

California, 1958: Surfers frustrated at calm seas attach roller skates to wooden boards for some "sidewalk surfing" instead. Fifty years later: the West Coast craze has spread worldwide. It captivates about 12 million young Americans annually. They're the kids in Vans sneakers who whoosh past on city streets astride their fiberglass boards on urethane wheels. Or they ride up walls in skate parks, orbiting like airborne dervishes. Some enthusiasts are psyched about skateboarding making the Olympics in 2012. Purists say it'll kill its soul.

648. Emily Dickinson's Greatest Poem: *Deathward bound.*

Death permeates her poetry, which has titles like "I Felt a Funeral in My Brain," "I Heard a Fly Buzz When I Died," and "My Life Closed

Twice Before Its Close," and such closing lines as "The liberty to die," "On the look of death," and "First chill, then stupor, then letting go." So it is a relief to know, from "I Never Saw a Moor," that Dickinson, who died in 1886, believed in heaven. Our choice of her finest poem is the one in which she depicts death not as a tyrant but a gentleman: "Because I could not stop for death / He kindly stopped for me; / The carriage held but just ourselves / And Immortality."

647. Newspapers: *Black and white and almost dead all over.* It's too early to bury them, but not to write a first draft of their obituary. Newspaper circulations are declining nationally, while the average age of readers is 55. This industry has been under siege from other media ever since Guglielmo Marconi took to the airwaves. While print journalism fended off challenges from radio, newsreels, magazines, television, and newsletters, it is basically on life support in the face of online forces. The Internet has stolen away classified advertising, once newspapers' cash cow, and wooed young readers to blogs and entertainment Web sites. Its immediacy and instant gratification mocks the 24-hour turnaround time of a daily newspaper. Still, we greet every morning with a cup of coffee and two newspapers delivered right to our door. And, for as long as our newspapers keep publishing, we celebrate this old fogey, civilized start to our days.

646. Soap Operas: *Life, love, and death in the afternoon.* It's amazing how little can happen in 30 or 60 minutes on a soap opera. In the 1980s Erica Kane, on *All My Children,* revealed she was about to start a disco. More than 200 episodes and innumerable love triangles, intrigues, and family crises later, she finally had the grand opening. Soap operas used to be for stay at homes and the unemployed. Recording lets us watch them anytime. Snobs may say these shows lack redeeming social value, but in a free-market economy, a product that thrives for 75 years must be giving people what they want. *Guiding Light* holds the longevity record. It started on the radio in 1937, switched to TV in 1952, but was switched off in 2009 for a heinous crime: low ratings.

645. Mesa Verde: *Native American architecture at its finest.* Eight hundred years ago the Anasazi built elaborate stone cities at the base of the Rocky Mountains in present-day Colorado. Around 1300 they abandoned them. In the 20th  century a rancher found their rooms "strewn with baskets and sandals, as if the residents had left only moments before," according to one account. They are perhaps the most spectacular, if equally mysterious, remnants of a lost American civilization.

644. *National Geographic*: *Vicarious tour guide.* Since 1888 this magazine, with its breathtaking photos and vivid accounts, has transported readers to far-flung territories, ocean depths, and limitless skies. Armchair adventurers can push through New Delhi's mobbed streets, eyeball a banded toe fish in coral reefs, or dodge a lunging leopard seal.

643. Banana Splits: *Gooey goodness.* Plain vanilla types may cringe, but banana splits are over-the-top delicious. The recipe is simple: split a banana lengthwise as a base. Top a scoop of chocolate ice cream with chocolate syrup, cover a scoop of strawberry ice cream with strawberry syrup, and smother vanilla ice cream in crushed pineapple. Squirt with a healthy serving of whipped cream. Sprinkle with chopped walnuts and maraschino cherries. The original treat sold at Tassel Pharmacy in Latrobe, Pennsylvania, for 10 cents in 1904.

642. The Baltimore Oriole: *The bird, not the team.* This American songbird was named for Lord Baltimore because the distinctive orange and black feathers were the colors of the Maryland colonizer's coat of arms. *Icterus galbula* can be found from Saskatchewan to the Atlantic Coast and as far south as Texas.

641. Weird Laws: *Legacy of grassroots democracy.* Our favorite is Kansas's stipulation that if two trains meet on the same track, neither shall proceed until the other has passed. Alabama bans fake moustaches that make people laugh in church, Florida prohibits unmarried women from parachuting on Sunday, and Massachusetts limits mourners at wakes to three sandwiches.

640. Rodeos: *Ride 'em cowboy.* The story goes that these ropin', wrestlin', and ridin' events began on July 4, 1869, when cowboys met in Deer Trail, Colorado, to compete at ranching chores. More than 600 rodeos now take place annually around the country, with professional rodeo cowboys traveling the circuit to vie for hefty purses. The classic event: saddle bronc riding. The goal: last an interminable eight seconds in the saddle astride a very angry, bucking horse.

639. The Amish: *A simpler way of life.* They settled in the heart of Lancaster County, Pennsylvania, in the 1730s. They reject modern amenities—like cars, televisions, electric appliances, and computers. A farming people at heart, their population has grown in the last two decades, now numbering about 227,000 and spreading their simpler lifestyle to communities in 28 U.S. states.

638. Carlsbad Caverns National Park: *Rock of ages.* Nature started making the 110 limestone caves in New Mexico's Guadeloupe Mountains 280 million years ago. We suggest spending a few hours in the largest, Carlsbad Cave, either on a relaxed tour or a strenuous rock climb. For the claustrophobic there are miles of trails aboveground.

637. Tabasco Sauce: *Fire in a 2 oz. bottle.* The red pepper condiment produced on Avery Island, Louisiana, and beloved worldwide, spices up practically any dish except ice cream. A friend splashes it on her morning bowl of Wheaties, milk, and fruit for a pick-me-up she swears beats coffee laced with brandy.

636. San Juan Islands: *Northwest havens.* Just a ferry ride north of the Washington mainland, these lush isles offer a temperate climate, picturesque harbors, and abundant fishing. Just ask Seattle residents who flock there to escape unending rain and traffic jams. The largest island, Orcas, has Mount Constitution, whose 2,409-foot-tall summit offers amazing views of Mount Rainier and Canada.

635. Joy of Cooking: *Recipes for success.* If you are only going to have one cookbook in your kitchen, buy this one. Its first edition, published in 1931, was written by homemaker Irma Rombauer during the depths of the Depression. Her 1,152 pages of common sense, cost-conscious recipes to feed our families resonate today.

634. Trail Running: *Soar ankles.* For getting in touch with the scenic wilderness and pushing a body to the limits of endurance, nothing beats hurtling up and down rock-strewn mountain paths. One of the best states for the sport is Georgia, where the 40-mile Pine Mountain Race is held every December and the winning time is often less than 6½ hours.

633. The Oregon Trail: *Showcase of the pioneer spirit.* In the 19th century the 2,170-mile trail was the pathway to the Pacific for hundreds of thousands of homesteaders, gold prospectors, fur traders, and other adventurers. Today, deep wagon ruts along remnants of the trail, which snakes across the Great Plains and the eastern Oregon desert, over the Blue Mountains and on to the Columbia River, remind hordes of 21st century hikers and history buffs of the courage of those long-ago settlers, 200,000 of whom died on the journey.

632. The Yosemite Series: *Through a lens starkly.* Photographer Ansel Adams (1902–1984) captured the beauty of the High Sierras in two-dimensional, black and white images suffused with light. *Bridalveil Falls, Cathedral Peak and Lake, Moonrise from Glacier Point,* to name just a few. These timeless photographs appear so often in calendars, posters, and books, they are imprinted in our collective psyche.

631. Turner Classic Movies: *Our cure for insomnia.* Can't sleep? Turn on this popular cable TV channel and nod off to the likes of Bette Davis, Van Johnson, and Gregory Peck. When media mogul Ted Turner bought and sold the MGM/UA movie studios in the 1980s, he held on to a library of more than 3,000 films,  which became the foundation for his TCM cable channel. The other movie channel, AMC, now airs its own programming and commercials, so we celebrate TCM's commercial-free format.

630. Winterthur: *Thanks, Hank!* The home, collections, and grounds of industrialist Henry Francis du Pont near Wilmington, Delaware, have become one of the world's most enchanting museums. The highlights are the rooms devoted to distinct periods of American furniture and designs.

629. Trader Joe's: *Destination for shoppers with champagne tastes and beer budgets.* Count us among the fans of the free coffee tastings and Two Buck Chuck, the cheap but decent wine that sells for as little as $1.99 a bottle. A kind of populist gourmet-food chain (with stores in more than two dozen states), Trader Joe's offers up treats like rack of lamb, lemon curd, Trader Josef's bagels, frozen crab cakes, organic sunflower oil, and orchids, all at cheapo prices.

628. Apollo Theater: *Venerable venue.* It started as a white burlesque house on West 125th Street in New York's Harlem in 1914, no black patrons allowed. Ironic that, because today the Apollo showcases mainly African-American talent. In the 75-plus years of its new incarnation, Apollo's headliners have included Smokey Robinson, Stevie Wonder, and Diana Ross, as well as Jamie Foxx, Mariah Carey, and Morrissey.

627. Ravinia: *Oldest outdoor music festival in North America.* There are other summer music festivals—Tanglewood in Massachusetts, the Monterey Jazz Festival in California—but Ravinia combines a unique blend of music genres. About 600,000 visitors gather annually at this former amusement park. Performers since this summer festival began in 1936 sound like a who's who of international music: rockers Janis Joplin and Frank Zappa, jazzman Louis Armstrong, cellist Yo-Yo Ma, tenor Luciano Pavarotti. About 25 miles from downtown Chicago, Ravinia is also the summer home of the Chicago Symphony Orchestra.

626. The Green Bay Packers: *In a league of its own.* This team, founded in 1919, has won a record 12 National Football League championships, including three Super Bowls, and is the most successful community-owned sports franchise in North America—a nonprofit corporation with about 112,000 shareholders.

625. Get-Rich-Quick Schemes: *Poker in real life.* The yearning for overnight riches is a universal human impulse. In America, ordinary folk have a better shot than people almost anywhere else at getting rich from a market insight and a bit of luck, like the middle-aged couple in Kansas City who concocted Greenies dog treats for canines with bad breath and made a fortune.

624. Food Stamps: *Licking hunger.* No longer considered a stigma of poverty, this federal aid has become a crucial support for low-wage earners, especially during tough economic times. Launched in 1939, it was renamed the Supplemental Nutrition Assistance Program in 2008. That same year it served 24.8 million low-income Americans each month, at a total cost of $37.5 billion.

623. *A Streetcar Named Desire*: *Landmark of our national stage.* Tennessee Williams wrote this Pulitzer winner in 1947. The plot pits delusional Blanche DuBois, who symbolizes the fading Old South, against her abusive brother-in-law, Stanley Kowalski, a member of the rising urban class. Their inevitable clash leads to Blanche's breakdown. The title may seem like a literary contrivance to explain how Blanche arrived in New Orleans, but the city actually had a Desire streetcar line from 1920 to 1948.

622. *Jeopardy!*: *TV entertainment for the intellectual elite.* The current version of *Jeopardy!* has been around since 1984, but this game show that attracts a smart, aging population actually debuted in 1964 with Art Fleming as host. The show made international headlines in 2004 when trivia buff Ken Jennings reigned as champion for 75 game days, winning more than $3 million. An American original, international versions of *Jeopardy!* abound. When you are traveling abroad, check out *Waagstuk!*, which airs in Belgium and the Netherlands, and *Rischiatutto!*, Italy's version.

621. Colonial Furniture: *Wood turned to gold.* An appraiser told Claire Weigand-Beckmann that the card table she bought at a yard sale for $25 was made by 18th Century Boston cabinetmakers at John Seymour & Son. It fetched $541,000 at auction. What's in your attic?

620. Homegrown Booze: *Saying thanks to our hard-drinking ancestors.* The fermentation of wine dates back to Neolithic times, the distillation of spirits to ancient China and Greece, and the brewing of beer to the civilizations of Sumeria and Babylonia (where the ration for high priests was five liters a day). So Americans can be justly proud of adding Kentucky bourbon, California wines, organic beers, hard apple cider, and moonshine whiskey to the liquor cabinet in a relative blink of a bloodshot eye.

619. Prohibition's Failure: *Lesson in legal restraint.* The constitutional ban on the sale of booze lasted just 14 years, from 1919 to 1933, and created more havoc than social uplift. Even so, we recommend restraint in the indulgence of alcoholic beverages—in moderation, of course.

618. *Birds of America*: *Capturing our feathered friends.* Nature lover John James Audubon's lively paintings of nearly 500 of the 700-some regularly occurring North American species depict these creatures in painstaking depth, color, and detail. In his book, first published in 1827, he took a new tack, noting birds' food and habitat preferences, movements, interactions, and behavior. Case in point: his Worm-eating Warblers show all aspects of the adult birds' plumage and, characteristically, how they forage on dead leaves.

617. Clark Art Institute: *Art and nature in harmony.* Tucked amid the bucolic Berkshire Mountains in western Massachusetts, the Sterling and Francine Clark Art Institute is one of the world's best kept secrets. This extensive 19th century art collection was built with the fortune Sterling Clark inherited from his grandfather, a founder of Singer Sewing Machine Co. The museum opened next to Williams College in Williamstown, Massachusetts, in 1955. Its 140 acres include intricate garden pathways that showcase the environs as one more canvas visitors can appreciate. We like Pierre-Auguste Renoir's buxom *Blonde Bather* and Edgar Degas's bronze *Little Dancer Aged Fourteen* decked out in a gauze tutu.

616. Piano Bars: *La la la, di da da.* True, they are not uniquely American. But they're more numerous and more fun in the U.S. Billy Joel's 1973 song about tickling the ivories in an obscure tavern—with its refrain, "Well, we're all in the mood for a melody, and you've got us feelin' all right"—makes us want to leave our computer right now and head for the nearest watering hole with live piano music.

615. Gas Station Attendants: *The pride of New Jersey and Oregon.* If you are a senior citizen, a pregnant woman, or just lazy, it is a pleasure to sit behind the wheel and let a professional fill the tank in the two states that ban self-service gasoline stations. There is no proof that motorists pay a premium for the service, according to an MIT study.

614. Prairie Dogs: *Rodents of the range.* Sociable and smart, with a sophisticated language of barks that identify predators, prairie dogs are also ecological assets, fertilizing grasslands and delivering rainwater to the water table through their tunnels. Farmers and ranchers detest them because they dine on their pastures and croplands; as a result, human encroachment and eradication programs have reduced their original habitat by 98 percent. However, a movement to preserve their colonies is gaining, uh, ground.

613. The Architecture of I. M. Pei: *Fusion of art and technology.* Best known for the 70-foot glass pyramid at the Louvre in Paris—a controversial project ultimately completed in 1989 and hailed as a masterpiece—the Chinese-American Pei has designed gasp-inducing buildings all over the world. In the United States

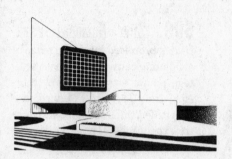

they include an addition to the National Gallery of Art in Washington, the John F. Kennedy Library in Boston, and the Rock and Roll Hall of Fame in Cleveland. Born in 1917, he was still making waves in his 90s.

612. The National Book Award: *Big win for worthy wordsmiths.* This recognition was established in 1950 by publishers, editors, writers, and critics to honor each year's best works in fiction, nonfiction, and poetry. The impetus was to get more Americans to read, but it didn't hurt if more books got sold too. Recent awards have gone to Tom Wolfe's *A Man in Full*, Shirley Hazzard's *The Great Fire*, and Joyce Carol Oates's *Blonde*.

611. Schlitterbahn Water Park: *The best splash.* Located in New Braunfels, Texas, this family-owned facility is the most popular water park in the United States. Schlitterbahn, which means "slippery road" in German, features the Master Blaster, a six-story-tall roller coaster with nine jets shooting water that is regularly voted the best water ride in America.

610. Cops on the Beat: *The thin blue line.* You don't want to mess with these gunslinging peacekeepers, and it's comforting to know the bad guys don't either. They put their lives on the line every day to keep us safe, and we put these lines on paper to salute them. More than 120 police officers, special agents, corrections officers, and parole and probation officers were killed in the line of duty in 2009. Happily, that figure was down 13 percent from the previous year.

609. Zion National Park: *Utah's promised land.* The maze of sandstone canyons and mesa vantage points attract so many tourists, the U.S. Park Service has sharpened the limits on backcountry expeditions in hope of keeping civilization from ruining Zion.

608. Political Paraphernalia: *Carnival of dumbed-down ideas.* Posters, bumper stickers, hats, balloons, and, of course, buttons, from the bland "I like Ike" (President Dwight D. Eisenhower's slogan in his 1952 and 1956 campaigns) and "All the Way With LBJ" (President Lyndon Johnson's pitch in 1966) to Texas politician J. J. "Jake" Pickle's pickle-shaped pin in the 1930s, all add to the fun. A California pub fastened a Barack Obama likeness to a beer tap in the 2008 "alection" ("ale"—get it?) campaign. Supporters of Sarah Palin, Obama opponent John McCain's choice for vice president, wore T-shirts that proclaimed: "I'm voting for Sarah . . . and that guy too."

607. Miniature Golf: *Putting to prominence.* About 27 percent of us play this, but minigolf is more than just a fun, family pastime. Trade associations and the pro circuit take it very seriously. It's achieved World Games status, and enthusiasts dream of the Olympics. Kitschy layouts with lava-spewing volcanoes are yielding to lush fairways, tricky doglegs, water hazards, and fake sand—much like the first U.S. minicourse in Pinehurst, North Carolina, in 1916.

606. The Golden-Winged Warbler: *Ugly duckling of songbirds.* There's nothing distinctive about this tiny slip of a drab gray bird except for the splash of yellow on its head and wings. And, then, it opens its mouth and sings.

605. Grits: *Glue that binds the South.* Grits are hominy ground into porridge, and we believe a serving tastes best with butter, maple syrup, and a topping of bacon bits. A liking for this southern comfort food is hard to acquire in adulthood, so we recommend that northerners who order a big breakfast in grits country tell the server, "I'll take the hash browns."

604. Crater Lake: *Deepest of the Deep.* Yes, Crater Lake in Oregon is the deepest lake in the world! Purists will aver that it is the ninth deepest, trailing no. 1 Lake Baikal in Siberia by a long shot. However, of lakes whose basins are entirely above sea level, Crater has the greatest average depth.

603. Marketing Co-ops: *Communal capitalism.* Americans are adept in this democratic commercial model. Consider biggies Ocean Spray, an agricultural cooperative owned by about 600 North American cranberry growers, and Florida's Natural, representing more than 1,000 citrus growers in central Florida. Scaling down, there's The Cheese Board in Berkeley, California, with only 17 owner-operators, each skilled in everything from recommending any of 250 cheeses to baking bread. They provide the equity, have voting rights, and split the profits.

602. Thanksgiving Dinner: *The once-a-year, all-you-can-eat gluttonous feast.* Turkey, gravy, stuffing, mashed potatoes, marshmallow sweet potato casserole, string bean casserole, cranberries, turnips, creamed onions, and corn bread,

followed by pumpkin, apple, pecan, chocolate cream, and lemon meringue pies, Pepto-Bismol, football, and, finally, a nap.

601. Children's Aid Society: *Helping needy kids.* Next time you think New York is a coldhearted place, think of its charity for disadvantaged youngsters. The society gives away 90 percent of the $100 million donated each year.

600. Elvis Sightings: *Tabloid grist with a messianic twist.* Isn't this country great? Not only did it produce the king of rock 'n' roll and, both before and after his death from an overdose of prescription drugs in his bathroom in 1971, a tidal wave of imitators. It also triggered a worldwide quasireligion in which the swivel-hipped crooner makes untold resurrection appearances. There have been thousands of apparitions—millions, some say—mostly in mundane spots like fast-food restaurants.

599. The Freedom Trail: *Revolutionary stroll.* Taking this 2.5-mile walk from Boston Common to Bunker Hill is a concrete way to grasp what shaped our national government. The red-brick route features 16 historic places, including the Old North Church and the site of the Boston Massacre. Bravo, Bostonians, for saving these structures from the wrecking ball in 1958.

598. MTV: *More than music.* Music Television (MTV) debuted in 1981 as a cable channel filled with nonstop music videos and transformed an auditory industry into a visual one. Somewhere along the way, MTV morphed into a teenage channel that featured music but emphasized entertainment. Media historians say the reality craze that still holds the U.S. television audience captive began in earnest with the broadcast of MTV's *The Real World* in 1992. Its travel show spinoff, *Road Rules,* popularized travel reality shows. In 1985, MTV launched VH1 for the fans who still wanted to zone out in front of music videos.

597. Do-It-Yourself Disasters: *Pumping money into the skilled labor economy.* The government should keep statistics on how many billions of dollars Americans pay plumbers, electricians, and auto mechanics to set right ham-handed attempts to fix toilets, appliances, and cars instead of calling the professional in the first place.

596. Advertising Home Runs: *Slogans that stick.* Funny, tantalizing, or in-your-face, these gems soared above the noise of TV commercials and the monotony of magazine ads to strike a chord in even the most strident critic of Madison Avenue's slippery ways. Here is a test: which products did our top 10 tout (answers below)? (a) I can't believe I ate the whole thing. (b) Where's the beef? (c) Zoom-Zoom. (d) Think outside the bun. (e) Let's build something together. (f) Does she, or doesn't she? (g) Diamonds are forever. (h) Just do it. (i) I'm Sharon: fly me. Answers: (a) Alka-Selzer. (b) Wendy's. (c) Mazda. (d) Taco Bell. (e) Lowe's. (f) Clairol Hair Color. (g) De Beers. (h) Nike. (i) National Airlines.

595. Think Tanks: *Opinions at large.* Research organizations such as Brookings Institution, Rand Corp., or Heritage Foundation may have their own political bents, but their often provocative findings invigorate the national debate. For example, Brookings brainpower represents more than 200 international experts in government and academia who provide research, policy recommendations, and analyses on a full range of public policy issues. Subjects of their thought-provoking papers can range from the importance of religion on Chinese society to ways of fixing the U.S. election system.

594. March Madness: *Basketball at its best.* Sixty-five of the toughest NCAA Division I college teams begin a three-week-long tournament in mid-March. Even more fun than the televised basketball

is the proliferation of illegal betting pools that clog up office copiers and keep workers preoccupied with the real reason they head to work in March.

593. Georgia Peaches: *Here comes the fuzz.* We were disappointed to learn that Georgia's peach production of 35,000 tons a year is third among states behind California and South Carolina. Yet we cling to the idea that a Georgia peach is still the best, almost 140 years after Samuel Rumph of Marshallville discovered a sweet new variety and named it Elberta, his wife's middle name. Why not Clara, her first name?

592. Levittown: *Pioneer of the postwar housing boom.* A planned community of several thousand Capes and ranch houses that was mass-produced on the potato fields of Long Island between 1947 and 1951, Levittown was heralded as the archetype of the new suburban dream. True, in its early years, it excluded people of color, singles, and the elderly, but it set the standard for providing affordable housing in a quasirural environment for young families.

591. Ultimate Frisbee: *A movable feat.* Mix soccer's athleticism, football's passes, and basketball's transitional aspects. Take two seven-member teams, give them a flying plastic disc to fight over, and let them loose. Invented in Maplewood, New Jersey, in 1968 by Columbia High School students, "Ultimate" stresses fair play. It has spread to more than 42 countries and boasts World Games status.

590. Heavy Metal: *Far out!* We're not talking the Captain and Tennille here, but the raspy, throbbing electric distortion sound of an American-born genre that traces its roots to the 1960s hard rock bands like Cream and Iron Butterfly. A special nod goes to WSOU-FM, the Seton Hall University radio station that has been celebrating heavy metal for more than 20 years. It was Metal Station of the Year for three years running.

589. The Big Game: *Being true to your school.* Harvard-Yale. Army-Navy. Duke-North Carolina. The game students and alums want to win most is the one against the hated, longtime rival, and the stands are filled. Pennsylvania schools Lehigh and Lafayette have been foes in football since 1884. In ice hockey, nothing beats the annual Beanpot—a tournament featuring Harvard, Boston University, Boston College, and Northeastern. We agree with ESPN and rank Ohio State v. Michigan the no. 1 college football rivalry, mostly because both teams are usually among the best in the country.

588. Mount Rainier National Park: *Throne of the North-west.* Rearing up 14,400 feet in full view of Seattle 54 miles away on a clear day (good luck!), the snow-cloaked mountain has 26 major glaciers and two volcanic craters. The park has hiking trails through old-growth forests, fields of wildflowers, glassy ponds, the 168-foot Narada Falls, verdant meadows, mineral hot springs, creeks, cliffs, and 1,000-year-old trees in the Grove of the Patriarchs.

587. Dollar Stores: *Where you get more bang for your buck.* You can buy a Christmas mug, wrapping paper, decorative napkins, an apple cutter, a spatula, and lots of other  really good things at these stores where everything costs a dollar. This national trend emerged from the old 5 & 10 cent stores. When five cents didn't even buy penny candy anymore, it was an idea whose time had come.

586. The Rio Grande: *The splash down south.* For 1,250 of its 1,865 miles the river is the boundary between the United States and Mexico, and the images that come to most minds are the shallows through which illegal immigrants try to come north. We prefer to dwell on the river's scenic wild side in the big bend country of West Texas, where rapids make for exciting rides between 1,000-foot cliffs.

585. The Pulitzer Prizes: *Celebration of the written word.*
Newspapers may be dying, but at least they are well written. We pat
ourselves on the back with these awards funded from the estate of
newspaper editor Joseph Pulitzer. Established in 1917, they celebrate aspects of daily journalism, playwriting, book and music publishing. The *New York Times* and its reporters have nabbed 101 of
these annual awards.

584. The Algonquin Hotel: *Home of America's most famous
literary club.* From 1919 to 1929 such luminaries as writer and wit
Dorothy Parker, humorist Robert Benchley, and playwright Robert E.
Sherwood lunched at the Algonquin (Manhattan's oldest operating
hotel, located on West 44th Street) to trade gossip and wisecracks.
They called themselves the Vicious Circle but newspapers chris-
tened them the Round Table. Drop by for a martini and a gander
at Natalie Ascencios's playful painting of the group in the Round
Table Room.

583. The Alternate Route: *Fresh classroom perspectives.* Pioneered in New Jersey in the 1980s, this solution to teacher short-
ages in public schools has evolved into a sophisticated model for
recruiting, training, and certifying people who have at least a bachelor's degree and want to teach. Now, instead of going the way of
traditional college education programs, there are 130 alternate
paths to certification in all 50 states and the District of Columbia.
Numbers taking this route have increased substantially since the
late 1990s. Nationally, about one-third of new hires are coming
through alternative routes to teacher certification.

582. *New York Times* Crossword Puzzle: *Mindless escape for
the sophisticated brain.* Most anyone can tackle Monday's crossword puzzle in the *Times*, but the word games get more difficult as
the week progresses. It takes years of practice to wrestle with the
weekend offerings.

581. White Castle Hamburgers: *Burp of a nation.* Clothed in oniony glory, the little burger with the big heartburn invented in 1921 is the Plymouth Rock of America's grab-and-gobble culture. Tiny holes in the meat let the steam soak through to create a unique texture. If you're not in one of the dozen or so states that have White Castle restaurants we feel sorry for you.

580. Memorable Movie Lines: *Why didn't I say that?* The American Film Institute's list of the 100 greatest movie quotes of all time is topped by "Frankly, my dear, I don't give a damn" (from *Gone With the Wind*, 1939). Other winners include: "I'm going to make him an offer he can't refuse" (*The Godfather*, 1972), "Toto, I've got a feeling we're not in Kansas anymore" (*The Wizard of Oz*, 1939), and "Here's looking at you, kid" (*Casablanca*, 1942). Our favorite is "Plastics," the advice given to Ben, played by Dustin Hoffman, in *The Graduate*. We also like "Go ahead, make my day" in the Dirty Harry movie *Sudden Impact*, produced in 1983.

579. The National Register of Historic Places: *Ones for this book.* A 1966 federal law authorizes the National Park Service to identify, evaluate, and protect America's historic and archeological sites. The register is the official list of these treasures, many of which we've cited in these pages. But there are lesser known gems like the Will Rogers Park Gardens and Arboretum, noted for landscape design, in Oklahoma City. And the Willa Cather Properties, Webster County, Nebraska, the settings for her fiction. Or the Big Duck, near Flanders, New York, which housed a poultry store in the 1930s and is now hailed as an architectural hallmark.

578. Art Institute of Chicago: *Children are welcome.* This midwestern museum invites young people to visit, and features family exhibitions designed to teach children to appreciate art. We especially like this suggestion: buy a group of postcards at the gift shop and then have the children find the paintings during their visit. Or, their Touch Gallery, which allows visitors a hands-on experience with sculpture.

577. Mount Washington: *Big East champion.* Puny by western standards, the tallest summit in the northeast is 6,288 feet above sea level in the White Mountains of New Hampshire. It is accessible to hikers, drivers, and passengers on a cog railway that opened in 1869, but the sudden changes in temperature can make the mountain dangerous.

576. Chinese Fortune Cookies: *Messages with bite.* After you've chewed the last of the cheap chow mein and egg rolls, it is always a treat to break apart those curved, hollowed-out wafers and read the platitudes printed on strips of paper inside. Samples: "Grant yourself a wish this year," "A warm smile shows a generous nature," and "You will inherit a large sum of money." Little known in China, they are handed out free by the billions at eateries in the United States.

575. Grand Central Terminal: *Beaux Arts destination.* A 1990s renovation restored entrepreneur Cornelius Vanderbilt's mid-Manhattan train depot to its former glory on opening day in 1913: melon-shaped chandeliers, a spectacular ceiling mural of the Mediterranean sky, two marble staircases connecting with the cavernous 80,000-square-foot Main Concourse. Every weekday now, more than half a million commuters ride the rails of this, the world's largest train station, with 44 platforms and 67 tracks.

574. eBay: *Online auctioneers.* In 1995, eBay founder Pierre Omidyar posted a broken laser pointer for auction to test his fledgling site. It sold for $14.83 to a collector of broken laser pointers, proving there's a market for everything. In 2007, eBay generated $7.7 billion in virtual auctions and sales.

573. The Bagel: *The doughnut's holier-than-thou cousin.* Invented by the Poles, appropriated by Jews because they could be baked quickly at the end of the Sabbath, these dense, chewy, yeasted-dough snacks have become an American culinary masterpiece. You have to go to the Big Apple to get the real thing, though. We love how you can order "an everything with nothing" (a bagel baked with sesame, garlic, poppy seed, onion, and salt but with nothing spread on it) or a "nothing with everything" (a plain bagel slathered with lox, cream cheese, and onion).

572. Name that Generation: *Pigeonholing the population.* Americans like to think of themselves as rugged individualists. Nevertheless, gotta love those demographers and marketers who need to divide us into six age groups demarcated by historical events. First, there's the dwindling GI generation, born from 1901 to 1924, and supposedly characterized by civic mindedness and cautious spending. Next, the Silent Generation, born from 1925 to 1945, and known as big spenders, followed by Baby Boomers, born between 1946 and 1964, with their "me first" attitude. Fourth is Generation X, birth years 1965-1980, and labeled the most misunderstood, scarred by divorce and single-parenting. Finally, there's Generation Y, born between 1981 and the present. They're like the GIs, only more positive. Who knows where Z will take us?

571. White Pass and Yukon Route Train Ride: *The railway built of gold.* Roller coasters are for sissies. This railroad, built during the Klondike Gold Rush in 1898, blasted through mountains to connect Skagway, Alaska, to the Yukon, 110 miles away. Today, the 40-mile ride over harrowing trestles and precarious bridges provides land excursions for hundreds of thousands of Alaskan cruise-ship passengers.

570. Model Railroad Clubs: Livin' la vida loca-motive. Gabbing about technical stuff. Buying and selling cool-looking historical items that actually move. Forging friendships. From the Palm Beach Model Railroaders to the Alaska Live Steamers in Wasilla, train lovers meet to talk about their collections and run them on hilly layouts. The annual train show in West Springfield, Massachusetts, attracts 24,000 enthusiasts. When he was a kid, Lawrence Schneeberger watched the 20th Century Limited rumble by. "I always wanted to ride on it but I never got the chance." In 2008, at age 100, he built a model version in his apartment in Burton, Michigan.

569. American Oratory: Thunderbolts out of the few. From Revolutionary war hero Nathan Hale's defiant cry on the gallows, "I only regret I have but one life to lose for my country" to George W. Bush's September 11 address proclaiming, "A great people has been moved to defend a great nation," the most famous actors on the stage of American history are the ones who matched the momentousness of the occasion with the eloquence of their words.

568. The Staten Island Ferry: Our best free boat ride. Some 60,000 passengers a day ride the ferries between St. George, Staten Island, and Whitehall Street in lower Manhattan. For most of them it's just a commute. For the rest of us the five-mile ride is a fantastic way to take in New York City's breathtaking skyline of skyscrapers and bridges. The ferry deck offers four-star views of the Statue of Liberty, Ellis Island, and New York Harbor.

567. Nickelodeon: First channel for children. Nickelodeon debuted in 1979 during the original explosion of cable channels. It

transformed children's programming by providing hours and hours of cartoons and adventure shows. Programs like *Today's Special, Belle and Sebastian, Double Dare, You Can't Do That on Television*, and *Salute Your Shorts* brought an edginess to children's TV and gave the 1980s generation of kids a new reason not to go out and play.

566. The Triple Crown of Racing: *Historic hoofbeats.* Even Americans who don't know a fetlock from a flintlock follow the annual attempt by three-year-old horses to win the Kentucky Derby, the Preakness, and the Belmont Stakes, accomplished 11 times, first by Sir Barton in 1919 and last by Affirmed in 1978.

565. After-Christmas Sales: *Giveaways disguised as economic transactions.* Bargains just don't get better than the thrift-shop prices you pay for the leftovers of holiday shopping madness. The quiet desperation of the sellers would be enough to silence the most ardent hagglers, if Americans haggled.

564. The Pledge of Allegiance: *Oath to the red, white, and blue.* In elementary school we learn the drill: hand over heart, we declare our fealty to the American flag, our Republic, and our "one nation under God, indivisible, with liberty and justice for all." The earliest version, just 23 words, appeared in 1892 in a youth magazine for students to chant during the 400th anniversary of Columbus's discovery of America. Although millions recited it for 50 more years, the pledge was made official by Congress only in 1942. Twelve years later it grew to 31 words, including the addition of "under God." That's the part critics assail as violating religion clauses in the First Amendment.

563. Off Broadway: *Where great theater is born.* The Fantasticks ran Off Broadway for 42 years. More typical is *Rent* or *Hair* or *Godspell* or *Doubt,* which all opened in New York City Off Broadway to much acclaim and eventually moved to Broadway. Off Broadway is defined as a New York City theater space with between 99 and 500 seats. It used to signify experimental-type productions, but today it is almost as established as the Broadway theater district. Now Off-Off-Broadway shows like the kind featured in New York's annual Fringe Festival is where you'll find the offbeat.

562. Majoring in Philosophy: *Education for its own sake.* This is a salute to the shrinking roll call of college kids who prefer the search for truth to the scramble for big bucks. Their study choice hearkens to the Middle Ages when the purpose of higher education was to answer big questions: is there a god, what is knowledge, do we exist, and what kind of cheap wine shall we drink as we discuss such stuff deep into the night?

561. Billy Graham Crusades: *Spreading the Gospel to the masses.* America's most famous evangelist preached to 215 million people in churches and stadiums over half a century, while 2.5 billion heard his broadcasts. Detractors scorned his message as feel-good theology, but he and his son, Franklin, who runs the Billy Graham Evangelistic Association, have done a pretty effective marketing job of getting the Word out.

560. *Our Town:* Play on words. One of the most popular American plays ever produced—first staged in 1938, it is always being performed *somewhere—Our Town* depicts a timeless tapestry of American family life without sets, forcing audiences to concentrate on the dialogue. They discover that playwright Thornton Wilder's masterpiece, while it transports them back to early 20th century America, reveals eternal truths about the human family.

559. Petition Drives: *Signing for a cause.* Whether protesting global warming or merely qualifying candidates for elections, petitions can be important democratic tools. One dramatic case: removal of Governor Gray Davis in California in 2003, followed by the election of Arnold Schwarzenegger—all traced to millions signing recall petitions.

558. Kingda Ka Roller Coaster: *59 seconds of terror.* Located at Six Flags Great Adventure in Jackson, New Jersey, this roller coaster zooms to 128 mph in 3.5 seconds. A hydraulic launch catapults riders 45 stories in the air before plunging to earth in a weightless spiral. We're a little skeptical but our kids tell us this is what they call fun.

557. The Smell of Honeysuckle: *The sweet scent of memory.* We encountered it as children, walking in a wood or meadow, and forever after the smell reminds us of the simple days of our youth. It's a pity the plant is a weed that harms other flora.

556. Motorcycle Jackets: *Machismo market.* Even accountants blossom into bad boys of the road when they zip up the leather. There's no need to hop on a Harley; the tight fit alone summons the ghosts of James Dean and Marlon Brando, ridin' on the wild side of the same skinned animal hide that makes couches so cozy.

555. Congressional Gold Medal of Honor: *Marks of gratitude for national greats.* Our highest and most distinguished civilian award, first presented in 1776, has gone to more than 100 recipients to honor singular acts of exceptional service and lifetime achievement. Congress approves each award, then commissions the U.S. Mint to design and create it. Honorees have included heavyweight boxing champ Joe Louis, the U.S. Marine Corps Navajo Code Talkers, and golf star Arnold Palmer.

554. **Sports Halls of Fame**: *Memorializing greatness.* Baseball fans head to Cooperstown, New York, football fans trek to Canton, Ohio, basketball fans flock to Springfield, Massachusetts, to honor their heroes. But don't stop there. There are sports halls of fame for tennis (Newport, Rhode Island), boxing (Canastota, New York), bowling (St. Louis), swimming (Fort Lauderdale), volleyball (Holyoke, Massachusetts), soccer (Oneonta, New York), lacrosse (Baltimore), golf (St. Augustine), hockey (Eveleth, Minnesota), wrestling (Stillwater, Oklahoma), and cycling (Somerville, New Jersey).

553. **Visiting Nurses**: *Homegrown Nightingales.* The idea dates to the 1880s in New York and Philadelphia, where free nursing care was provided to the sickest and poorest who would otherwise have gone untreated. Today, the Visiting Nurse Associations of America represents nonprofit, community-based home health care and hospice providers who work with more than 4 million low-income elderly and disabled people. "Society benefits when health care is provided in the least costly and most comforting setting—most often the home," said pioneer public health nurse Lillian Wald. With skyrocketing health care bills, it can also curb hospital costs.

552. **Mount Vernon**: *Presidential retreat.* George Washington's serene plantation lies 16 miles from the nation's capital on the Potomac River. This 18th century working farm and gristmill is America's most visited home. Its 500 acres include woods, meadows, and gardens designed by Washington himself. As he wrote, "No estate in United America is more pleasantly situated than this."

551. **The Grand Teton National Park**: *Big-breasted beauty.* The Tetons (which derive their name from the French word for nipples) in northwest Wyoming include the 13,770-foot Grand Teton and eight other soaring peaks. From 200 miles of trails, visitors soak up views of the alpine terrain, shimmering lakes, sagebush flats, wildflower fields, and wildlife that includes 300 species of birds and the rare Rubber boa snake.

550. Kentucky Blue Grass: *Another immigrant makes it big in America.* Settlers from Europe brought over the hardy *Poa pratensis*, and it has become the dominant strain of grass in places in America with plenty of rain. The name comes from the blue flowers that sprout when it reaches its natural full height of two or three feet.

549. Country Fairs: *Rural rollicks.* From the hippie-inspired Oregon Country Fair outside Eugene that features jugglers, clowns, storytellers, dancers, musicians, puppet shows, comedy skits, crafts, food, and, at least in the old days, cavorting bare-breasted women, to old-fashioned county fairs with their queens and princesses, steer and hog shows, woodworking exhibits, chili cook-offs, apple pie contests, quilt shows, horse-and-pony pulls, parades, rodeos, Ferris wheels, and ostrich races, they are just plain fun.

548. Ticker Tape Parades: *Triumphant whiteouts.* On special occasions since 1886, snowstorms of shredded paper—originally culled from ticker tape machines—have swirled down from office windows in lower Manhattan's "Canyon of Heroes" to celebrate politicians, explorers, athletes, pontiffs, and war heroes. Recently so honored: the New York Yankees for their 27th World Series championship in October 2009. They were showered with about 50 tons of confetti and shredded paper.

547. Johns Hopkins Hospital: *Medicine with an international flavor.* When railroad magnate Johns Hopkins died in 1873, he left money to establish a hospital/university that trained competent

and caring medical staff. It was the first U.S. hospital linked to a university. Today, Johns Hopkins in downtown Baltimore enjoys an international reputation as a hospital that welcomes patients from around the globe. Its Web site is available in 10 languages and its doctors annually treat patients from more than 100 countries.

546. *The Simpsons*: *The funhouse mirror.* Historians in 1,000 years will learn from episodes of *The Simpsons* what America was all about in the decades before and after the turn of the 21st Century, just as we study Chaucer to understand old England. In the meantime, we're laughing our heads off at Homer, Marge, Bart, and Co.

545. Colorado: *A great place to visit, but you'd probably want to live there too.* With the Rockies in its western half and the plains in the east, it is one of our most beautiful states. With its ranches, factories, gold mines, gas and oil fields, forests, and the Wall Street of the West in Denver, it is one of the wealthiest. It was the first to grant women the right to vote (Wyoming beat it to the punch, but as a territory). It has the lowest obesity rate, and contains the remotest county outside Alaska. It is closest to the clouds, and you can go to its border with Arizona, New Mexico, and Utah and put your foot in four states at once.

544. Biscuits: *Savory sides.* To most of the world, biscuits are a hard, baked food. To Americans, they're tender and flaky bread products. The difference: adding baking powder and pats of butter to the dough. Southerners smother theirs with gravy. Northerners slather theirs with butter and jam. Whatever your preference, try stopping at just one.

543. Forsythia: *Spring's first burst of glory.* For 50 weeks a year forsythia looks like a weed that needs pruning. It justifies its existence in early spring when it bursts into bloom in spectacular yellow. It's so ubiquitous that people assume it's an American plant, but like the cherry blossoms, forsythia originated in Asia.

542. Garage Bands: *Rock around the block.* The "garage rock" genre started in the 1960s with songs like "Louie, Louie." Now it's also a term for start-up bands, often in a garage or basement or attic. Pity the parents of the drummers. Their homes are usually where it's happening. Our favorite all-time garage band? Bill and Ted's Wyld Stallyns, dude.

541. Our Overseas Possessions: *Unreal real estate.* Aside from the 50 states, the federal government presides over unincorporated organized territories like Guam, Puerto Rico, and the U.S. Virgin Islands; unincorporated unorganized territories (American Samoa and uninhabited islands like the Midways); incorporated organized territories (for example, the Palmyra atoll, an archipelago of 50 tiny islands owned by the Nature Conservatory); and unincorporated unorganized territories (including overseas military bases). These far-flung spots enrich our culture, provide scenic travel destinations, and give strategic advantages to our military planners.

540. Farmers Markets: *Urban harvests.* These markets number more than 4,600 nationwide, up nearly 3,000 since 1994. They're popular because they bring the country to city dwellers who clamor for locally grown, fresh produce as a tastier and healthier alternative to chemical-laced, store-bought fare. Considering recent massive recalls of contaminated spinach and tomatoes from the nation's supermarkets, consumers are paying much more attention to their food and its origins. Farmers markets give them the chance to chat with the producers. Add neighborhood musicians and face-painters, and shopping becomes a street festival.

539. Cardinals: *The scarlet feather.* We're going out on a limb here to say the cardinal is America's most beautiful bird. The males are a bright scarlet, and even the duller females sport a startling red beak. The only negative: the cardinal is an early riser and starts its day with a melodic song loud enough to wake the dead.

538. National Gallery of Art: *The place to be seen.* France has the Louvre, Spain the Prado. With his gift of art and money, tycoon Andrew Mellon made possible the 1941 opening of America's National Gallery of Art in Washington, D.C. The thousands of works range from sculptures by ancient Greeks to oils by Harlem-born abstract expressionist Norman Lewis. The 40 paintings by Gilbert Stuart always attract a crowd.

537. Delicate Arch: *What the wind blew in.* You have to hike a mile and a half off the beaten path for this breathtaking sandstone arch in the middle of Arches National Park in Utah. It is worth the trek.

536. The Gateway Arch: *Stupendous stainless steel stopping point.* Completed in 1965, it is an engineering marvel and an architectural wonder. You can take a ride to the top of the 630-foot-tall monument in St. Louis that bills itself as the entryway to the West for magnificent views in either direction. Don't miss it on your next cross-country trip.

535. Frederick Remington's Art: *Westward ho!* Eastern-born Remington (1861–1909) discovered his muse out West in the 1880s. His action scenes of soldiers, cowboys, and Native Americans inform our image of frontier life to this day. Remington's bronzes, among the best American sculptures, include *The Bronco Buster*. A casting of this classic symbol of the American West is displayed in the Oval Office.

534. Brownies: *Blissful squares.* However you like 'em, fudgy or cakey, these chocolate desserts are ubiquitous staples at fairs, barbecues, and ladies' luncheons. Some swear brownies were invented in the early 1900s by resourceful Maine home economist Mildred Schrumpf, after her chocolate cake collapsed in the oven.

533. Game Shows: *Play along at home.* You can find out if you are a savvy consumer (*The Price is Right*), if you are smarter than a fifth grader, or if you want to be a millionaire, just by watching game shows. We like the *Cash Cab* that drives around New York City, picks up unsuspecting fares, and then quizzes them en route to their destination. Three wrong answers, they are thrown out mid-ride. It's all good!

532. American Kidney Fund: *Dollars for dialysis.* It collects more than $80 million a year to help 75,000 kidney disease patients pay for treatments that keep them alive. The fund holds administrative expenses to four percent of collections, one of the best ratios among major health charities.

531. Ghost Towns of the West: *Relics of frontier days.* The saloons, whorehouses, and general stores are boarded up and crumbling now (except for the tourist haunts that capitalize on Wild West shows), and the mud streets are silent, but once upon a time these desert outposts bustled with cowboys, gunslingers, gold prospectors, and other adventurers. There are thousands of them scattered throughout the West, bearing names like Bootstrap (Nevada), Shakespeare (New Mexico), and Humptown (Idaho).

530. Radio City Music Hall: *Art Deco heaven.* The world's largest indoor theater, this New York City wonder is a masterpiece of American modernist design, with marble and gold foil played against

Bakelite, aluminum, and cork. Perfect acoustics and uninterrupted sight lines mean all 6,200 seats are good. Completed in 1932, it's best known as home to the Rockettes, the famous precision dancers whose high-kick line is a show-stopper in the annual Christmas Spectacular.

529. The Endless Yard Sale: *654 miles of someone else's junk.*
Once a year in August, vendors and amateur sellers line U.S. Highway 127 for the mother of all yard sales. It snakes through five states: Ohio, Kentucky, Alabama, Georgia, and Tennessee. You can buy lots of *really good stuff* and take time out to visit some of the Civil War battlefields that line the route.

528. The *Star Trek* Phenomenon: *It has the power.* Since the original 79 episodes on TV from 1966 to 1969, it has become a cultural cornucopia, spilling out new series and movies, trekkie revivals and fan fests, and phrases like "Beam me up," "I can't hold her, Captain," and our favorite, "I'm a doctor, Jim, not a magician."

527. Arizona's Petrified Forest: *The plant kingdom's ancient ruins.* This landscape of 225-million-year-old fossilized wood is almost too enchanting: Despite security guards, fences, and warning signs, visitors make off with 12 tons of the trees' remains every year.

526. *Appalachian Spring*: *Harmonies from the heartland.* A perennial favorite since its premiere in 1944, the simple melodies of composer Aaron Copland's Pulitzer Prize–winning ballet score cloak a masterpiece. As biographer Howard Pollock puts it, Copland's musical method, replete with jazzy meters, playful rhythms, odd beats, and sudden harmonic shifts, "suggests immediate parallels with American history and society," invoking our westward expansion and our social fabric, with people of widely diverse backgrounds uniting in a common national purpose. An emotional

highpoint: the sweet melody based on the traditional Shaker song, "Simple Gifts."

525. iPhones: *The ultimate pocket gadget.* Two years after these handheld computers/phones debuted in 2007, more than 1 billion applications had been downloaded. Sure, iPhones do what all cell phones do—text and chat—but they also play music and surf the Net. Then there are thousands of quirky "apps" that let you calculate your blood/alcohol levels, charter a private jet online, translate foreign languages, view traffic snarls on your route to work, distort your voice, and, our favorite, make your phone fart instead of ring. We also are quite impressed with the Google Droid phone, a new kid on the market.

524. Whistler's Mother: *No way to treat a lady.* James Whistler's 1871 portrait of his mother, titled *Arrangement in Grey and Black,* was a breakthrough study in form and color. Now it is the often parodied image of the mythic old American woman, garbed in black, humorless and serene. It might be the best known 19th century painting by an American, but you must go to the Musée d'Orsay in Paris to see the original.

523. Mega Churches: *Super-sizing Christianity.* Critics say these behemoths that attract 2,000 or more worshippers to services can sometimes care more about growth than substance. Still, with their gyms, bookstores, and food courts, and the lifelines they throw to the desperate (like Las Vegas gam-

blers), they meet practical as well as spiritual needs. Plus, notoriously church-averse kids seem to love them, and they are catching on abroad.

522. Barbershop Quartets: *Pitch-perfect pipes.* Uniquely American, these foursomes feature unaccompanied close-harmony singing and expanded sound from distinctive chord structure. We associate this style with white singers, but in the late 19th century it was African-American men, working in barbering jobs, who pioneered these groups. Today, the Barbershop Harmony Society, a major force started in 1938, boasts over 30,000 male members in 800 U.S. chapters. Women do their harmonizing with the Sweet Adelines.

521. Café du Monde, New Orleans: *Heaven's version of Dunkin' Donuts.* Follow the coffee aroma to the Café du Monde, the gateway to the outdoor French Market that dates back to the days of the Choctaw Indians. The coffee, flavored with the smooth taste of chicory, and the *beignets*—a French-style doughnut that smears your clothes and face with powdered sugar—are how we imagine breakfast in the afterlife.

520. Vacation Shares: *A house divided.* Pooling the rent money for a cottage at the beach, the lake, or in ski country is a way to get away for less, hang out with pals, and learn rules of etiquette. Such as never stub out a cigarette in a sliver of cold pizza crust. You could be ruining someone's breakfast.

519. The National Cathedral: *A capital site to see.* This English Gothic masterpiece, completed in 1990, 83 years after construction began, is the sixth-largest cathedral in the world, with a tower that commands the highest point in Washington, D.C. Visitors gawk at its mosaics, gargoyles, and 200-plus stained-glass windows. Music lovers delight in the sounds of its world-famous choirs and Great Organ, which has 10,647 pipes. Funerals were held there for Presidents Dwight Eisenhower and Ronald Reagan.

518. The U.S. Naval Academy: *Molding midshipmen.* Future officers in the Navy or Marine Corps receive state-of-the-art academic and professional training at this Annapolis, Maryland, institution, established in 1845. Although they will be leaders in a naval force that is the largest and most modern in the world, their vessels will still fly the First Navy Jack, the famous *Don't Tread on Me* standard—reinstituted for our fighting ships after the War on Terror began in 2001.

517. 19th Amendment: *Pivotal stepping-stone to real democracy.* How can you talk about democracy when less than half your population can vote? Women in the United States today take this privilege, granted by the 19th amendment in 1920, seriously. Census data show there are more women than men in the United States and the women have a higher rate of voting. President Obama benefited from this phenomenon: 53 percent of all voters were women, and 56 percent of all women voted for him.

516. New York's Operas: *Where the fat ladies sing.* The Metropolitan Opera, founded in 1883 and the showcase for stars from tenor Enrico Caruso to soprano Renée Fleming, is the nation's premiere opera company. It's also the world's most widely heard, thanks to weekly broadcasts. We're big fans too, of the upstart New York City Opera—if an institution founded in 1943 can be considered an upstart. Twenty-nine new operas debuted there and it was the first company to provide supertitles with English interpretations.

515. Santa Claus, Indiana: *St. Nick's mailbox.* More than half a million kids send him letters every year, and volunteer "elves" in this hamlet of 2,200 answer them all. Sure, Bethlehem, Pennsylvania, Noel, Missouri, North Pole, Alaska, and the 140 municipalities in the U.S. with "Christmas" in their names all evoke the Yuletide season, but none conjures up visions of sugarplums in so many little heads as this place does.

514. Veterans Organizations: *Serving those who served.* Some 50 groups champion the interests of our nation's 25 million veterans, and rightly so. The 2.2 million member Veterans of Foreign Wars, for one, helped establish the Veterans Administration, pushed for GI bills, a national cemetery system, and compensation for vets exposed to Agent Orange or diagnosed with Gulf War syndrome.

513. Mammoth Cave: *Stalactites and stalagmites unite!* This longest cave in the world has more than 365 explored underground miles. Kentucky's Mammoth Cave National Park was established in 1941 to protect the labyrinth. Hardy tourists in hiking boots can opt for the six-hour, five-mile, Wild Cave tour that promises a taste of real spelunking. Gloves and knee pads are recommended.

512. The Washington Monument: *Casting a shadow on the White House.* This is not about that 555⅛-foot shaft of marble on the capital mall. It is homage to that invisible tower of traditio ns, created by the first President, that sits alongside his successors. The framers of the Constitution created a strong executive branch because they knew that the ramrod-straight hero of the Revolution would be the first to hold the office and they trusted him to establish the proper precedents. More than two centuries later his character and ideals survive in the model of what a President should

be: courageous, a consensus builder, wary of partisanship. And Presidents have learned from his example that they don't have to be the best educated person in their administrations. They can lean on the wisdom of Jeffersons, Hamiltons, Adamses, and Madisons. All a President needs to know is who to listen to, and when.

511. Warhol's Cans: *Chicken soup for American pop culture.* These 32 canvases by pop-art icon Andy Warhol (who died in 1987 at age 58) depict mundanity itself, yet somehow transform the ordinary into the extraordinary. One of the silkscreen paintings sold for $11.8 million. That makes the struggling artist in all of us want to shout, "Yes, we *can!*"

510. Fund-raisers: *Bartering for betterment.* Nobody knows how much is raised annually in this country through car washes or sales of entertainment books and wrapping paper. Trust us, the total take is in the many millions. It's a tribute, then, to American generosity that people still dig deep to pay for team equipment, class trips to Disney World, or new church vans. Just avoid what one Florida group did to raise money for its safe-driving cause: a fund-raiser in a martini bar.

509. Diamond Head: *Aloha, big boys!* In the 1800s, British sailors named this site on the southeast coast of Oahu in Hawaii for its calcite crystals in lava rock, glittering in the sunlight. Actually, this is the crater of a volcano, extinct for 150,000 years, some 3,520 feet in diameter with a 760-foot summit. Tourists treasure this attraction for its hiking trail: the strenuous, steep climb delivers a spectacular ocean view.

508. Facebook: *Social networking Web site used by everybody who is anybody.* Facebook connects hoards of "friends" worldwide.

Founded by a student at Harvard University in 2004, Facebook allows users to share their individual virtual "page" of photos and personal information with friends. It replaces outdated college yearbooks by linking potential friends and rooting out the six degrees of separation that connect us all.

507. Anthologies of Old American Music: *Where have you gone Buddy Bolden?* Jazz and popular music from the early days of recording are on CDs thanks to music archivists. Our favorite is the two-volume Best of Jabbo Smith, the trumpeter. The Holy Grail is the wax cylinder rumored to have been made by Buddy Bolden's band, considered by many to be the first jazz ensemble.

506. St. Augustine: *First among firsts.* Lying on Florida's First Coast, it is the nation's oldest city (founded in 1565), home to its oldest port, location of its first Catholic mass, site of its first Underground Railroad safe house and of its first major Greek settlement, and producer of the only trickle of water that has a credible claim to be the legendary Fountain of Youth sought by Spanish explorer Juan Ponce de Leon.

505. To Hell with the Joneses: *Consumer culture rebellion.* Your neighbors buy a BMW, hire a landscape architect, take a cruise, and beautify their kids with nose jobs and dental caps. You resist the temptation to keep up with the bourgeois strivers and earn the right to feel smug when the bank forecloses on their house.

504. The Varying Sizes of States: *Geographic aid for dummies.* Most of the world's 195 nations are compared in the press to one of the states, beginning with Iran and Mongolia, both Alaska-sized, going through Texas-sized France and Ohio-sized Iceland down to Rhode-Island-sized Dubai, leaving a handful at the top that have to be likened to the whole damn country (India is one-third the size, Russia twice) and a sprinkling at the bottom, like Lichtenstein, that suffer the indignity of being likened to cities such as Dayton, Ohio, or the Vatican, which could fit on a golf course.

503. Children's Literature: *Books that last forever.* *The Very Hungry Caterpillar; Pat the Bunny; The Little Engine that Could; The Golden Egg Book; Goodnight, Moon; Black Beauty; Little Women; Toyon: A Dog of the North and His People; Nancy and Plum; The Giving Tree; Louis the Fish; Alexander and the Terrible, Horrible, No Good, Very Bad Day; Charlotte's Web; The Boxcar Children; Mr. Popper's Penguins; Ramona Quimby, Age 8; A Wrinkle in Time; Little House on the Prairie; Bridge to Terabithia; Tales of a Fourth Grade Nothing; How to Eat Fried Worms; Where the Wild Things Are; Frog and Toad; Sarah, Plain and Tall; Miss Nelson Is Missing;* and *The Story of Ferdinand.*

502. The Frick: *Venus rising from the slag heap.* The New York mansion coal and steel magnate Henry Clay Frick built in 1913 houses the best small art collection in America. The ambience brings visitors back to the Age of Innocence. And oh those artists: Rembrandt, El Greco, Holbein, Whistler, Stuart, Cellini, Vermeer, Van Dyke, and Romney, to name just a few.

501. The Sundance Film Festival: *Indie Nirvana.* The most prestigious event of its kind, this antidote to big studio Hollywood, annually showcases more than 200 new, independently made  documentaries and feature films. Launched in 1978 by actor/director Robert Redford, the festival draws 50,000 to screenings in 12 theaters in the Sundance, Utah, area. Popular flicks first shown here include *The March of the Penguins, Little Miss Sunshine,* and *Precious.*

1001. Procrastinators Club of America: *Perfect meeting place for people who aren't joiners.* Whoops. This was going to be no. 938 on our list of the 1,000 things to love about America, but we only got around to writing it now. The first mention we found of this group was a *New York Times* story in 1980 that quoted its president as saying it had 600,000 members, but only 3,500 had officially signed up. It hasn't made the news in recent years, for whatever reason. Whoever is in charge of the club's Web site has been putting off adding anything of much interest, though we like the slogan "productivity is overrated" and are intrigued by the link to "seven ways to procrastinate for better results," which we must read someday.

500. Federal Funds for College: *Investment in our future.* The federal government recognizes its obligation to help educate our next generation through several programs, most especially the $1.4 billion Work Study program and the $19.3 billion Pell Grants. Work Study funds paychecks for students who hold down jobs on their campuses. The Pell program provides grants to low-income students. For many high school graduates, college would be an idle dream without these financial lifelines.

499. The Florida Keys: *Laid back in the sun.* These islands below the Florida mainland have escaped much of the awful development that is destroying the Atlantic shores. Key West is a haven for gays, nonconformists, and partygoers of all orientations. The best attraction is Nobel prize–winning novelist Ernest Hemingway's house, now a museum.

498. *American Gothic*: *Remembrance of things past.* It's the most instantly recognizable 20th century American painting. This 1930 oil by Grant Wood memorializes farming life—father and spinster daughter armed with a pitchfork to till their family-run farm. In truth, Wood asked his sister and his dentist to pose for the painting. You can find the original at the Art Institute of Chicago.

497. Comic Book Collections: *Kids' caches.* Heroes of this low-brow literary genre—cartoon strips with simple plots and simpler dialogue—range from Superman and Wonder Woman in the golden era of the 1930s, '40s, and '50s, to Doktor Sleepless and TV spinoff RoboCop in our own day. Television and the Internet have dented but not destroyed comics' appeal to wide-eyed youngsters, to say nothing of their dreamy-eyed elders. Sales continue to be robust, topping 100 million copies a year.

496. Lost Comic Book Collections: *The same kids' caches, gone in a flash.* Ah, it would have been worth a lot of cash if only Mom hadn't thrown my Action Comics in the trash. Swapping such tales of woe ranks second only to bemoaning missed real-estate deals in the pantheon of middle-class America's financial might-have-beens. Look for inspiration in the 2004 auction of 400 comic books dating back to the 1930s and '40s that an elderly widow named Irene Ford Henschel found in bags under beds in her home off a dirt road in South Wichita, Kansas. Collectors paid more than $100,000 for the trove.

495. Utah: *The best contender against Hawaii for the title "the nation's most beautiful state."* The natural splendors of this vast and sparsely populated land blow away the most jaded travelers. It is home to five national parks—count 'em!—hugs a couple of others in Colorado and Nevada, and provides succor to tourist-averse campers in the isolated stretches of the Grand Canyon's northern rim. It boasts some of the most awe-inspiring peaks of the Rocky Mountains. Its other attractions, from gut-wrenchingly gorgeous river gorges, unexplored cliffs, salt flats, and sunset skies of pink, gold, and purple, have been immortalized in Zane Grey's novels and Ansel Adams's photography. You gotta give the Mormons credit: they chose a great location.

494. Regional Accents: *One nation, different vowels.* With our transient population and the neutral language of mass media, you'd think we'd all sound the same. Ain't so. Many experts say that American local dialects are more pronounced than ever in some places, especially cities. Nobody's certain why, but we shout, "Hallelujah. Let's hear it for cultural variety." So whether "youse" come from New Yawk or "y'all" hail from Nawlins, "pahk ya cah in Hahvahd Yahd in Bahsten," eat "chawclat" in Ohio, or sail your "bewt" in San Fran Bay, you're sweet music to our ears.

493. West Point: *The Army's Harvard*.
Britain's Duke of Wellington said of the defeat of Napoleon's army in 1815, "The battle of Waterloo was won on the playing fields of Eton." Shiloh, Gettysburg, Argonne Forest, Normandy, and Baghdad were decided at the U.S. Military Academy, where our U.S. generals—Grant, Lee, Pershing, MacArthur, Eisenhower, and Schwarzkopf—graduated. Rebecca Halstead, an Iraq War commander, was the first alumna to become a general. Spooky poet Edgar Allen Poe enrolled in 1830 but got the boot after one semester. There's a story, probably made up, that he wore a cartridge belt, gun, and nothing else to a drill.

492. Maple Syrup: *The sweet side of sap*.
Sugar: bad. Honey: good. Maple syrup: delectable. You tap the maple trees, boil the water off the goo, and put the resulting syrup in jars. Or you buy a bottle of the stuff in Stop & Shop. Either way, pour liberal amounts on pancakes, waffles, French toast, and ice cream, or just take a direct slug from the bottle. Delicious, and a good source of manganese and zinc.

491. Break Dancing: *Kinetic contortions*.
This improvisational street dance grew from the hip hop movement of New York City's African-American and Hispanic communities in the 1970s. It mixes dance, martial arts, and gymnastics performed to electro or hip hop music. Note the robotic routines, glides, one-hand handstands,

even head spins. We've loved them in Michael Jackson's music videos and films like *Wild Style*.

490. Lake Superior: *World's largest freshwater surface.* Gichigami, which means big water, is the Ojibwa Indian name for the largest of the five great lakes. Michigan, Wisconsin, and three Canadian provinces surround it. For superior recreational fun year-round, head for Bayfield, Wisconsin, gateway to the Apostle Islands and their historic lighthouses.

489. The Appalachian Trail: *Hikers' heaven, hikers' hell.* "The experience was one that I either actually hated or I numbed myself to it," declared Bill Bryson in his 1998 best-selling book *A Walk in the Woods* about puffing along a big chunk of the 2,175-mile pathway from Springer Mountain, Georgia, to Mount Katahdin, Maine. We feel his pain—in our brief bursts along sections of the trail, we encountered coiled black snakes, got lost in bogs, and clambered over fields of rocks. We also tasted exhilaration: breathing in the pure and silent air, exchanging yarns with "thru hikers" testing the trail's entire length, and, once, stumbling on the remains of a 1957 Studebaker in a forgotten glen.

488. Recycling: *Goodbye to all that garbage.* Americans recycle about 80 million tons of trash annually, which is about 32 percent of all our garbage. That figure might sound low, but it's double what it was 20 years ago. Also on the plus side, 75 percent of adults recycle garbage at home. It's become a national pastime: we buy products that are made from recycled waste; environmentally friendly cleaning products are au courant; we bring fabric bags to the supermarket; we sponsor computer recycling efforts; we buy energy-efficient appliances, hybrid cars, and fluorescent lightbulbs. The "zero waste" movement is gaining ground and

landfills are closing. After the creation of Earth Day in 1970, the green movement took decades to gain momentum, but it's finally reached a point in the American psyche that if we don't recycle and go green, we at least feel guilty about it.

487. Mobiles: *Movable marvels.* Thanks to American artist Alexander Calder (1898-1976), we are drawn to the abstract design of metal, steel rods, and wood, powered by motors, wind or water, and dubbed the "mobile." Check out his stunning 76-foot-long metal mobile hanging from the atrium roof in Washington's National Gallery.

486. Manatees: *Wrinkled elephants of the ocean.* You gotta love these guys. They weigh about 1,500 pounds, eat 100 pounds of food daily, and are as ugly as sin. They eat, rest, and travel (sounds good, huh?). There's something mesmerizing about watching these aquatic mammals munch vegetation and ponderously plod underwater. They are on the endangered list as "vulnerable to extinction" mostly because of collisions with boat propellers in the warm Florida waters where they congregate.

485. Walk-in Closets: *The conquest of inner space.* Old European apartments have charm, but nothing but armoires to hang clothes. Befitting a country with grand vistas, America's walk-in closets have forests of dresses and suits, vast valleys of shelves lined with shirts and sweaters and mountains of shoes. Even better is a walk-*through* closet. When the Associated Press reported that swindler Bernie Madoff's prison cell was the size of a walk-in closet, we were outraged. That big?

484. Courthouse Squares: *Nostalgia for a simpler way of life.* Old men in overalls still whittle sticks, chew tobackee, and gossip on benches in the shadow of the columns. A farmers' market sprouts up on the lawn every Saturday morning, just as it always has. And the Main Street shops, a minute's walk away, are still open for business—small-town America's show of defiance against the spirit-crushing mall culture that reigns in the suburbs.

483. **Fly Fishing**: *Water Zen.* Since the 1992 film *A River Runs Through It*, fly fishing has been luring many more devotees to U.S. waters. This sport uses artificial flies to hook fish, usually trout and salmon. Enthusiasts come to understand that when you marry solitude with natural beauty, time slips away. Then, ultimately, it's just you and the fish.

482. **Vocational Schools**: *Career incubators.* These workshops train students with little aptitude or appetite for academic study in blue-collar skills like carpentry, auto repair, and hairdressing. Recently, they have also embraced fields like health care and the food industry.

481. **Bob Dylan's Philosophy**: *Music for the changin' times.* Dylan's wisdom dwindled as he aged, but lucky for us he was productive enough in his formative years to last a lifetime. Dylan, the darling of the 1960s New York City coffeehouse scene, fueled the folk music genre, politicized our music, and encouraged us to live authentic lives. He chastised our parents, chided rotten politicians, and denigrated capitalist pursuit of wealth. His prognostications hold true, even today. The answer to just about any question you can ask, my friend, is still blowin' in the wind. Our favorite Dylan: *Subterranean Homesick Blues* and *Positively 4th Street*.

480. Artwork of Andrew Wyeth: *For-profit fusion of talent and hype.* Wyeth's masterpiece (or is it simply his best-known painting? Either way, it is an American icon) is *Christina's World*, the 1948 depiction of a young woman crawling across a farm field toward a hilltop house. His most ambitious venture was the Helga series, 247 paintings and drawings executed over 15 years of a strong-boned woman with braided hair in various, mostly nude, postures. He was still painting until shortly before his death in 2009 at the age of 91. Critics have called Wyeth's oeuvres corn-pone Americana, and branded the artist himself as a high-class huckster. The broader public adores him, and always will.

479. Organic Farming: *Cropland with class.* Organic farming shuns synthetic pesticides, fertilizers, feed additives, and growth regulators for crop rotation, composting, and biological pest control. It is no longer fringe but increasingly mainstream. This healthier approach represents one of the fastest-growing segments of U.S. agriculture. The number of organic farms in the United States more than doubled from 1992 to 2005, to 8,500 from 3,600. Also, the land under their cultivation more than quadrupled, to 4.1 million acres from 935,000 acres.

478. New Year's Eve in Times Square: *The TV show, not the real thing.* The smartest people in America at midnight every December 31 are the ones watching from home as the ball descends to the cheers of a million revelers in New York who would freeze if they weren't packed in like penguins.

477. The Global Warming Debate: *Fire and ice revisited.* The rhetoric gets heated, with true believers painting a greenhouse catastrophe of flooded coastal cities, mass species extinctions, and spreading deserts, and skeptics pooh-poohing their alarms as scare-mongering by socialist utopians. A consensus is emerging, however, that the earth's temperature has probably risen by at least one degree Fahrenheit over the past century, with man-made carbon emissions a major culprit. The question becomes how to deal with the consequences. That debate will be with us until the Arctic sea ice melts or we correct the problem. Let's hope poet Robert Frost was wrong in his conclusion that while ice might destroy the world, "From what I've tasted of desire / I hold with those who favor fire."

476. Food and Drug Administration: *The ingestion police.* If you chew it, eat it, drink it, swallow it, inject it, or rub it all over, chances are the FDA has passed judgment on it. Writer Upton Sinclair's sickening 1906 exposé of the Chicago meat-packing industry, *The Jungle*, is generally credited with galvanizing the government into action. All this makes America one of the safest countries in the world to pop pills or pig out.

475. The Music of Cole Porter: *Night and day, he is the one.* From the jazz age to the sophisticated '60s, Cole Porter wrote music and words to more than 1,000 songs, and dozens are standards. The clever lyrics keep the songs modern, and even performances of Porter's obscure ditties—like, "They All Fall in Love," are worth watching on YouTube. Besides "Night and Day," there are "Anything Goes," "Let's Do It," "I've Got You Under My Skin," "Brush Up Your Shakespeare," and "You're the Top," which displayed his genius for rhymes. "You're the top! You're an Arrow collar. You're the top! You're a Coolidge dollar. You're the nimble tread, of the feet of Fred Astaire. You're an O'Neill drama, you're Whistler's mama! You're camembert."

474. Western Wear: *Cowpoke couture.* For men, jeans with a studded leather belt, a bold-colored shirt with a black string tie, and a 10-gallon hat. For women, a denim skirt and jacket, braided horsehair bracelets, and a sexy  straw hat. For both sexes, rawhide boots. They wear 'em for real out West, for show in the East. Either way, they are pure Americana.

473. Route 66: *Memory two-laner.* Established in 1926, The Mother Road, as it's called, testifies to our independent spirit. Popularized by *The Grapes of Wrath*, a hit song, and a 1960s TV series, Route 66 links small-town America with major cities. It connects Chicago to Los Angeles, running through eight states and three time zones. Interstates now have trumped it. And souvenir hunters have swiped its signs. Still, romantics can explore most of it, riding some 2,000 miles past old villages, trading posts, filling stations, even the first McDonald's.

472. Action Figures: *Make-believe for boys.* For centuries, girls have called a lifelike toy with movable arms and legs a "doll." But if you're a boy, it's an action figure. The first one, GI Joe, was marketed by Hasbro in 1964. Most memorable are the Teenage Mutant Ninja Turtles that have been in stores for 20-plus years.

471. Federal Bankruptcy Law: *Starting a new chapter in your life.* Up to 2 million individuals, and tens of thousands of businesses, escape their crushing debt every year by filing for legal protection against creditors. The quickest route is Chapter 7 of the bankruptcy code, but Chapter 11 and Chapter 13 are also in much demand, and there is Chapter 9 for towns and cities, Chapter 12 for farmers and Chapter 15 for international cases. Sometimes you just have to go for broke.

470. New Deal Art: *What's in your post office?* In the Great Depression of the 1930s, the Works Progress Administration and Treasury Department paid 5,000 artists to create more than 225,000 paintings and sculptures. The best known are the murals in public buildings. For a good sampling, visit the 25 post offices in Oklahoma that have these works—many depicting Indians—by artists who earned $23.50 a week and were glad to get it.

469. The Texas Rangers: *All-purpose posse.* Created in 1834 when Texas was part of Mexico, the Rangers have fought Indians, spied on Mexican troops, put down riots, battled Union soldiers, investigated murders, stood up to the Ku Klux Klan, rounded up whiskey smugglers, and tracked down outlaws. Still in active service, they are the oldest state law enforcers in the United States.

468. Rhythm and Blues: *Earns our R-e-s-p-e-c-t.* Just think "I Can't Stop Loving You" by Ray Charles, or The Temptations' "Sugar Pie, Honey Bunch," or Aretha Franklin's "Respect." R&B covers a wide-ranging genre of popular music created by African Americans in the late 1940s and early 1950s. These days it's viewed as a modern version of soul and funk-influenced pop music. That said, we groove on singers Mary J. Blige and Patti LaBelle.

467. Wikipedia: *Everyman's online encyclopedia.* We used to trek to the library to research in weighty encyclopedia like *World Book* or the *Encyclopedia Britannica.* Then along came this bright American idea to ask the citizens of the world to create an encyclopedia using "wiki" technology that allows users to write and edit collaboratively. The result is more than 2 million footnoted entries in English. Other articles are available in hundreds of languages. School children, beware! While Wikipedia can give you a general idea of important topics, there's a lot of puffed up fluff amid the facts.

466. Maine: *Too scenic and cold for crime.* We're guessing that people who huddle by hearths in winter and breathe free among the forests and along the rocky coasts year-around don't have time to tangle with the law. Maine's rate of violent offenses has been among the states' lowest for a century, and the most recent national census statistics pegged Maine's 1.6 murders per 10,000 residents dead last among the states.

465. William and Elizabeth: *The comeback kids.* William had a 96-year run in the Social Security Administration's top 10 "Most Popular Baby Names" but fell off it in 1976, only to resurface early this century. In the same period that started in 1880, Elizabeth had a great ride until 1925, went into hibernation for decades, came back in 1980, stuck around through 2001, and has been popping on and off the list ever since.

464. Cranberries: *Bog bounty.* Native Americans showed starving pilgrims how to make cranberries palatable, thus their presence on Thanksgiving tables everywhere. Rich in antioxidants and fiber, the juice of these tart berries is said to ward off bladder infections. Down the hatch now!

463. Sardi's: *Sketchy favorite.* Located at the center of the universe, a.k.a. Manhattan's theater district, Sardi's has been feeding everybody from the glitterati to out-of-work actors and actresses to gawking tourists for close to nine decades. Customers love the 700 caricatures of famous people that adorn the walls, the Old World wait staff, and the lively atmosphere as much as they relish the cuisine, which is much more *haute* than local snobs admit.

462. UCLA Medical Center: *The best in the West.* When you are old and sick, move to L.A. Renamed the Ronald Reagan UCLA Medical Center and relocated to a new state-of-the-art facility, this Los Angeles hospital ranks no. 1 in the nation in geriatric care.

461. The Hudson River School: *Light and atmosphere on canvas.* A group of mid-19th century American artists painted scenes of the Hudson River Valley, the Catskill, Berkshire, and White mountains, and lands to the West in a romantic, sometimes theatrical way. The public loved them for it and so do we. Our favorite of the groundbreaking Luminist Movement: Frederic Edwin Church's dramatic *The Icebergs*.

460. *Leaves of Grass*: **Democracy in verse.** "Song of Myself," the longest poem in this 12-verse collection first published in 1855, is considered the most thoroughly democratic poem in world literature and a landmark distillation of the American identity. In it, poet Walt Whitman identifies with all manner of men and their conditions ("Of hue and caste am I, of every rank and religion."). To this day, the CollegeBoard touts Whitman's masterpiece as recommended reading for college-bound.

459. *Folk Art*: **Home sweet home.** Country scenes, simple figures, little proportion or perspective, pure colors, and strong details. These are characteristics of American Primitive, a style exemplified by the works of Grandma Moses (1860-1961), an elderly farm wife who taught herself to paint after arthritis crippled her for embroidery. Her paintings of late 19th and early 20th century scenes might as well be embroidery. A favorite is *Black Horses* (1942), with its verdant valley views.

458. *Deodorant*: **Nicety for our noses.** Okay, okay. Americans are too preoccupied with smelling nice, but given the choice of a crowded bus on a hot June day in Paris or New York, we'll take our chances with the New York crowd. The deodorant industry in the U.S. tops world consumption at $1.9 billion annually.

457. *American Dentistry*: **Top drill.** Most people fear dentists. Ask the estimated 167,000 teeth doctors practicing in the United States. Okay, having gurgling plastic tubes shoved in your mouth is no picnic, but consider this: our dentist's office displays brilliant artwork, skylights show

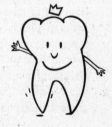

billowing clouds, the music is soothing, and the equipment cutting edge. That beats our dentist encounters in Europe—with our dated gold crowns to prove it.

456. Our Ingenuity: *Solving life's little challenges.* The country that brought the world dental floss, tea bags, cotton swabs, and the Zamboni is proud of its people's innate ability to problem solve. We radically improved the quality of our life by purchasing a handy little gadget that automatically waters the Christmas tree. Our Chia pet has a healthy head of sprouts; our Ginsu knife still cuts through any surface. If only our "Clapper" would keep the lights on when we sneeze, life would be perfect.

455. Colonial Portraits: *Windows into a world lit by candle.* The best early American painters studied in England, and most of their subjects were the rich and famous, like Revolutionary War hero Paul Revere, whose circa 1768 portrait by John Singleton Copley is in the Boston Museum of Fine Arts. The George Washington portrait Charles Wilson Peale painted before the Revolution hangs in the chapel at Washington and Lee University in Lexington, Virginia. Only a few examples from the 17th century survive, among them two oils in the Worcester Art Museum in Massachusetts by an anonymous artist known as "the Freake Limner" after the name of the family of proper Bostonians he painted.

454. Our Special Relationship with Israel: *Diplomatic line in the Middle East sand.* The United States hasn't wavered in its commitment to the Jewish state since President Truman recognized it hours after its existence was proclaimed on May 14, 1948. By his action (which he didn't bother to report in advance to State Department bigwigs), Truman forged an alliance with an islet of democracy in a sea of corrupt dictatorships. The decision to become Israel's first and truest friend has paid strategic dividends, and while it has inflamed the Arab street, it has also made us the key to any eventual peace settlement in the region.

453. PTA: *Children's champion.* Founded in 1897 by two Washington, D.C., mothers concerned about child safety, the National Parent Teacher Association is our largest volunteer, child-advocacy group with more than 5 million volunteers in 25,000 local units. Over the years, the PTA has been instrumental in promoting kindergarten, child labor laws, a public health service, hot lunches, a juvenile justice system, and mandatory immunization.

452. Food Network: *Fertilizer for couch potatoes.* Millions of Americans are armchair chefs. We watch others prepare meals that we will never attempt. Our favorites on this cable network: *Iron Chef America* and any show with chef Bobby Flay.

451. A Summer Weekend Down the Jersey Shore: *Hoi polloi heaven.* The migration starts on sticky Fridays in the New York–Philadelphia megalopolis, and even if their course to the ocean is northward, folks say they are going "down" the shore. At this 130-mile multicultural festival from Keansburg to Cape May Point, everyone has a favorite beach, a favorite bar, and two favorite restaurants—the one they go to and the one they would go to if the line were shorter. Swells stay in their mansions in Deal or Avalon. Singles jam group homes in Manasquan and Belmar, while Long Branch and Wildwood are full of families. There are subs, Philly cheese steaks, clams, crabs, and tubs of Bud on ice. The hot breeze carries the scent of sweat, sunscreen, and grilling burgers across a landscape of wooden decks. Then the traffic builds on the Garden State Parkway as the crowd slouches home to face Monday.

450. Alaska Polar Bears: *Classy carnivores.* They have to contend with global warming, oil drilling, and a shrinking habitat, but the creatures are making their last stand in the United States with aplomb. Experts say

they will still be roaming our northernmost state for decades—or longer, if we get the hang of this ecological thing.

449. *Nighthawks*: *Urban isolation in oils.* Edward Hopper's 1942 painting shows a waiter at a counter serving three customers. It's late night, but the diner is illuminated by fluorescent lights that cast an eerie glow on the sidewalk outside. The contrasts of light against dark and the menacing atmosphere hint of Film Noir. Hopper gave us an indelible image of mid-20th century America.

448. Grandma's Remedies: *Feel-good advice from the old folks.* Call them junk science, superstition, or just plain crazy, some of these do-it-yourself treatments actually work. Or they serve as a psychological boost that makes you think they work. A couple of our favorites: treating toenail fungus by soaking feet in apple cider vinegar or Vicks VapoRub; or, heating a jar of olive oil, dropping in a penny, tightening the lid, and applying the bottom of the jar to aching muscles. Many grandmas swear that chewing on raw garlic helps cure a host of ailments, and the kindlier ones permit the patient to munch a chocolate bar chaser.

447. License Plate Slogans: *Marketing mélange on the motorways.* Our favorite is New Hampshire's "Live Free or Die." We like the squabble between North Carolina's "First in Flight" invocation of the Wright Brothers at Kitty Hawk and Ohio's "Birthplace of Aviation" reference to the aircraft's construction in their Dayton bicycle shop. Then there is South Carolina's "Smiling Faces" versus South Dakota's "Great Faces," an allusion to Mount Rushmore. The District of Columbia makes a political statement: "Taxation without Representation." Others get right to the point (Idaho's "Potatoes"), indulge in word play (Louisiana's "LoUiSiAna," Utah's "Greatest Snow on Earth"), or are just plain cute (Georgia's " . . . On My Mind"). Wyoming stands (drives?) alone, with no motto at all.

446. Stanford University: *Center of cutting-edge curricula.* Founded in 1885 in Palo Alto by railroad magnate Leland Stanford and wife, Jane, this is one of the world's leading research and teaching institutions. Known for its law, medicine, business, and education schools, it's also neighbor to northern California's "Silicon Valley," home of Yahoo!, Google, and Hewlett-Packard—companies started and led by Stanford alumni and faculty.

445. *Monday Night Football*: *Tackling prime time.* Professional football was a Sunday afternoon affair until the ABC network brought it to Mondays in 1970. *MNF* successfully married sports and entertainment with witty and knowledgeable commentary by the likes of Howard Cosell, John Madden, and Tony Kornheiser. The broadcast has always welcomed celebrities to the broadcast booth. The most interesting pairing was in 1973 when future President Ronald Reagan explained the rules of the game to Beatle John Lennon. In 2005, *MNF* moved to ESPN, where it regularly breaks records for cable viewers, luring more than 18 million fans for some matchups.

444. Bubble Gum: *The flavor of childhood.* Walter Diemer invented it in 1928 by adding natural latex to the mix at the Fleer Chewing Gum Company. They called it Dubble Bubble. As long as Americans have teeth, they are forever blowing bubbles.

443. Pennsylvania: *Cradle of Liberty.* William Penn, the first colonial governor, preached it. The Declaration of Independence, signed in Philadelphia, demanded it. The Constitution, drafted in Philadelphia, enshrined it. The Battle of Gettysburg, the turning point in the Civil War, paid for it in blood. Pennsylvania is famous for steel and potato chips, for the nation's first oil well and first zoo, for its only bachelor President and for luminaries, native or transplanted, like Founding Father Ben Franklin, woodsman Daniel Boone, literary genius Edgar Allen Poe, and President Dwight D. Eisenhower. Its state dog, the Great Dane, alas, has lost out in stature to its pipsqueak rival, the weather-forecasting groundhog, Punxsutawney Phil.

442. Death Valley: *Extreme tourism.* The hottest, driest, lowest place in North America, this California-Nevada desert has sand dunes, snowcapped mountains, craggy canyons and 3 million acres of stone wilderness. The valley is a long basin 282 feet below sea level, hemmed by steep mountain ranges. Summer temperatures can top 120 degrees by day in the shade, 90 at night.

441. Showers: *The American standard for bathing.* More Americans prefer a long shower to a relaxing hot bath. Elsewhere around the globe, bathroom renovators are still trying to figure out how to get more than a dribble out of one of those awful handheld models.

440. The Works of Norman Rockwell: *Chronicler of our lost innocence.* His 322 *Saturday Evening Post* magazine covers from 1916 to 1963 depicted Americans in all their wholesomeness—desired or real. The kitsch police may scowl, but most of us feel uplifted, amused, or nostalgic when looking at such Rockwell illustrations as the 1943 renderings of Roosevelt's Four Freedoms, the 1947 Thanksgiving Dinner, or the 1958 image of a boy lowering his trousers as a doctor prepares an injection.

439. Going to Church: *Weekly retreat from the woes of the world.* About 40 percent of Americans attend worship services regularly (or claim to). Most attend to praise God. Some go to escape loneliness, others to network, and still others for the free food at coffee hour. Whatever their motives, they enjoy a safe harbor from everything from fear of eternal damnation to cell phone blabbers.

438. *The Invisible Man*: Black-and-white narrative.
This novel by Ralph Ellison, written in 1952 when Jim Crow reigned, chronicles the misadventures of a nameless black man to reveal the absurdities of racism in America and, more broadly, the blindness of the powerful for the underdog. With its narrative sweep and dark humor, it is considered a work of genius that puts its author above even James Baldwin and Richard Wright in the triumvirate of the great African-American writers of the mid-20th century.

437. Breakfast Cereals: Making us regular folks.
The average American consumes about 160 bowls of cereal a year. These crunchy grains we eat to start the day were developed to counteract gastro-intestinal disorders common in the late 1800s, when Americans breakfasted on heavy pork and beef, but no fiber. That's why Kellogg developed Corn Flakes, and Post, Grape-Nuts and Post Toasties. So clean your bowl for the sake of your bowels.

436. Wheat Fields: Amber waves of grain.
We Americans each consume 137 pounds of wheat annually—that's a lot of bread, pasta, and vodka. The United States is the fourth largest wheat producer in the world, behind the European Union, China, and India. Those beautiful acres that sway in the autumn breeze are a staple crop throughout the northern Plains and Midwest states.

435. Columbia River: Something to gush about.
The 1,200-mile torrent through Washington and Oregon that carried explorers Lewis and Clark to the end of their transcontinental journey in 1805 sends more water into the Pacific than any other river and is North America's top producer of hydroelectric power.

434. Bandannas: *All-purpose head gear.* Bandits in the Old West covered their faces with these colorful cotton squares. Today, rednecks and hippies alike flaunt them. Farmers, firefighters, and foot soldiers don them to protect themselves against dust and fumes. Kids make fashion statements with them at rock concerts. Street gangs use them to identify themselves. You can wrap them around the top of your head or your neck (or your dog's). You can buy them for as little as 30 cents.

433. Head Start: *A leg up early.* Created in 1965 as an outgrowth of President Lyndon B. Johnson's War on Poverty, Head Start is the country's longest-running, school-readiness program. Its education, health, and nutrition services target low-income children and their families. More than 25 million preschoolers have benefited. Studies show that by spring of their kindergarten year, Head Start graduates are even with national counterparts in reading and writing and nearly so in math and vocabulary.

432. Bruce Springsteen Concerts: *The Boss rules.* Does this guy ever take a night off? Springsteen packed Madison Square Garden for 10 concerts in 2000 and then filled the 80,000-seat Giants Stadium for 10 consecutive shows in 2003. He closed the concert series in the old Giants Stadium in 2009 with a week of sold out concerts. He actually likes touring! His E Street Sirius radio channel replays recorded concerts daily. About 400,000 rocked at his Obama inaugural concert in January 2009. A few weeks later his 13-minute miniconcert at halftime of the Super Bowl drew about 98 million U.S. viewers. Hardcore fans can sometimes find him jamming at Asbury Park's The Stone Pony, the bar that his E Street Band first called home.

431. Big Box Stores: *Markets for customers who live large.* We're not talking about Wal-Mart and Home Depot, but stores like Costco, BJs, and Sam's Club, which have limited selections of products in bulk and always have a shipment of something new and intriguing. Like 400-piece tool kits or gallon jugs of mango salsa. In the Costco checkout line one morning we admired the low price of the 30 packages of lamb chops in another customer's cart. "They'll be more expensive tonight," he said, handing us a card to his restaurant.

430. Delta Queen: *Floating landmark.* One of the last of the great steamships, the *Queen* was built in California in the 1920s but plied the Mississippi and connected rivers with its trademark red paddle for decades, well into the 21st century. Famous passengers included three Presidents, a Supreme Court Justice, and outsized personalities like boxer George Foreman and singer Tammy Wynette. Last we heard, it was a floating boutique hotel off Chattanooga, Tennessee. It is a national historic landmark, but then, so was the steamer *Avalon*, today renamed the *Belle of Louisville,* which one of us used to ride up and down the Ohio River as a kid.

429. Boston Marathon: *26 miles 385 yards of torture.* It's the world's oldest annual marathon. Inspired by the success of the first modern Olympic Games, the Boston Marathon started in 1897 with 15 runners. Grudging admiration goes to the estimated 22,000 wackadoodles who annually cross the finish line. It amazes us that there's a wait list to participate because so many people want to put their bodies through this torture.

428. The Associated Press: *The truth never gets old.* The cooperative that four New York newspapers started as a pool for Mexican War dispatches in 1846 is now the world's largest producer of up-to-the-minute print and broadcast news, financial data, photos, and video. As the media descends into a swamp peopled by self-important bigmouths, biased blowhards, and know-nothing bloggers, we hope the AP's 3,000 journalists keep digging for journalism gold: the facts.

427. Monticello: *Thomas Jefferson's monument to himself.* The house and gardens the author of the Declaration of Independence and third President designed and built on his 5,000 acre plantation near Charlottesville, Virginia, are windows into the mind of America's quirky Founding Father. He compared his Palladian style home favorably to Paris: "Tho' there is less wealth there, there is more freedom, more ease, and less misery." We wonder if his 187 slaves agreed.

426. Kentucky Bourbon: *Spirited rejoinder to scotch and vodka.* Recognized by Congress as our best native brew, it was consumed in large quantities by Ulysses S. Grant. It is produced mostly from corn. Tired of those snobbish wine tastings in California vineyards? Hit the Kentucky Bourbon Trail for a stronger kick, or show up at the Kentucky Bourbon Festival in September. Americans gave the hooch its name in gratitude for French help in winning our independence.

425. CNN: *Global eye.* In 1980, Turner Broadcasting System, Inc. launched Cable News Network, the first 24-hour, all-news television network. Transmitting via satellite to cable systems, the Atlanta-based company revolutionized broadcasting. Viewers could watch breaking news like the first Persian Gulf War in 1991 and terrorist attacks on September 11. Now owned by Time Warner, CNN reaches an estimated 93 million U.S. households and its global reach through CNN International is huge, airing in more than 212 countries and territories.

424. Frank Sinatra Music: *He did it his way.* Ol' Blue Eyes sure knew how to croon a tune. He died in 1998 at age 82, but his songs live on in elevators everywhere. Paul Anka penned the lyrics to "My Way" with Sinatra in mind. Our favorites include "All the Way," "Send in the Clowns," and "Fools Rush In."

423. The Missouri River: *The country's longest.* It starts in Montana, winds through the Dakotas, Nebraska, and Iowa, and empties after 2,300-plus miles into the Mississippi at St. Louis. Paddling its length through fast water, wide reservoirs, and the heavily trafficked lower reaches is a feat more rewarding than hiking the Appalachian Trail—"a magnum-caliber opportunity for adventure," says David Miller, who wrote the definitive guide to the river, *The Complete Paddler.*

422. Political Cartoons: *Art with attitude.* We loved David Horsey's "The Nightmare Team," which showed an annoyed Barak Obama clasping hands in the air with a grinning Hillary Clinton as a red-nosed Bill Clinton barged his way between them. Herbert Block ("Herblock") reigned supreme in the 20th century, penning caricatures of the high and mighty, like his depiction of President Nixon hiding behind a flag during the Watergate cover-up. Benjamin Franklin created one of the first political cartoons 1754, a snake cut into eight pieces with the caption "Join or Die," an appeal for colonial unity in the French and Indian War.

421. St. Patrick's Cathedral:
Catholic America's pride. In 1853, Archbishop John Hughes exhorted his meager following in New York City to build a grand, Gothic cathedral "for the glory of Almighty God." Lambasted as "Hughes's Folly," the site was then in the boonies, but Hughes envisioned it at the heart of the future metropolis. He was right. Today, Gothic St. Patrick's, with all its majesty, welcomes Presidents,

pontiffs, the pious, and even atheistic aesthetes through its Fifth Avenue portals.

420. Mac and Cheese: *Comfort casserole.* Originally a southern staple, this classic dish is a welcome standby up North now too. Some say gourmet Thomas Jefferson created it using a macaroni mold he brought back from Paris. We do know as our third President, Jefferson had it served at the White House in 1802. But please, no mixes here. Just macaroni in a cheese-based white sauce, topped by more cheese, then buttered crumbs.

419. Cancer Hospitals: *Centers of hope.* The two best cancer treatment/research facilities in the world are located in the United States. Memorial Sloan-Kettering Cancer Center on the East Side of Manhattan attracts patients from around the globe. Similarly, the M.D. Anderson Cancer Center on the Houston campus of the University of Texas enjoys international fame. They regularly battle each other for first place in the world.

418. Bloomberg Terminals: *Financiers' favorite toy.* For investors, traders, bankers, corporate managers, analysts, and regulators, the numbers-crunching machines are indispensable, even addictive. It's sort of like the Google of the financial world: anything you want to know about companies and markets, whether real-time information or historical trends, you can call up on the screen in an instant. The 300,000 or so terminals, which cost subscribers $1,500 and up a month, are also a cash machine for New York City Mayor Michael Bloomberg and his political ambitions.

417. Square Dancing: *The reel deal.* The roots are in the country dances that colonial immigrants brought from Europe, followed by anything-goes barn dances—the kind that end in hilarious drunken brawls in cowboy movies. The mannered modern version, known as western square dancing and featuring professional callers, developed in the 1930s and is more like an orchestrated minuet than a hoedown. The first bow to your partners, promenade, do-si-do, and allemande left are the beginning of a night of polite family fun.

416. Photo Treasures: *Gallery of the American saga.* The flag raising at Iwo Jima, a sailor kissing a nurse in Times Square at World War II's end, Einstein sticking his tongue out, Marilyn Monroe holding down her billowing skirt, refugees scrambling up a ladder from a Saigon rooftop into a helicopter. These and other classics form a visual panorama of nearly two centuries of our history.

415. Light-Frame Construction: *The rest of the world wood if it could.* Most countries use materials like masonry, steel, and timber to build houses and other small structures, while the dominant technique in the U.S. is frames stabilized by studs, ceiling joists, and rafters. They cost less, at least if there is an abundant supply of wood nearby.

414. Gospel Music: *"The Good News" songs.* This predominantly American phenomenon, descended from Christian hymns and Negro spirituals of the 19th century, has spread throughout the world. Gospel music embraces everything from "How Great Thou Art" to "Swing Low, Sweet Chariot," but the chief message stays the same: Christ's redemptive love. A recent favorite: Mercy Me's "Finally Home."

413. Wildlife Gone Wild: *They're back!* The bald eagle, symbol of American might, almost went extinct in the Lower 48, but because of conservation efforts now numbers 10,000 breeding pairs. Dozens of other species, from alligators to timber wolves to the Atlantic piping plover, a tiny bird that breeds on East Coast beaches, have returned from the brink. Still others, like coyotes and wild turkeys, that were a rare sight not long ago, are now everywhere, the former gobbling up puppies and the latter chasing postal workers.

412. Cleveland Clinic: *The Midwest's medical magnet.* U.S. *News & World Report* ranks this the best cardiac care center in the United States. More than 3.2 million people are treated here annually. A teaching hospital, it also provides medical and surgical training for health professionals. If you are going to have heart trouble in the Midwest, it's a good idea to get yourself to Cleveland.

411. Madison, Wisconsin: *Middle America's Oxford.* The capital of Wisconsin and the home of the state university's main campus finishes high each year in rankings of top towns for education, earnings, eco-friendliness, and recreation facilities. The only blemish is the cold—below 20 degrees on the average day from December 1 to March 1.

410. Home Schooling: *Freedom of voice.* Parents who don't like what the schools are (or are not) teaching have the option of taking on the job themselves—and a huge and growing number do. The 2 million kids from kindergarten through high school who learn their readin', writin', and 'rithmetic at home outperform their counterparts at both public and private schools in test scores, college graduation rates, and even spelling bees, though some say some fall behind in the art of interacting well with others.

409. Garage Sales: *Down-home markets.* Americans are forever relocating, and before they do, they sell, sell, sell. Want a used sofa, old baby clothes, an almost-new grill? These and more go for a song at garage sales. They're great for making money, eliminating stuff, and socializing. But no early birds, please!

408. Reality Game Shows: *TV's crazy craze.* Just how far will you go to win a million dollars? Would you travel to a remote jungle, starve yourself, dine on creepy crawlers, and play silly obstacle course games? *Survivor* is for you. If you are overweight, you can strip down to a sports bra and shorts and weigh in on national television on *The Biggest Loser*. Folks who enjoy travel can race around the world, bungee-jumping or hang-gliding off cliffs in *The Amazing Race*. Those who think reality game shows are just a passing fancy should note they dominated the first decade of the millennium. There seems to be no dearth of bizarre ideas to push the limits of an already crazy format.

407. Unofficial Town Web Sites: *All the news not fit to print.*

These online sources of news and information are starting to replace local newspapers for news that isn't filtered by self-interested politicians. The best of these sites, like the one for Saugus, Massachusetts, have a mix of public information, minutes, and transcripts of town meetings, plus forums for unfettered discussion. The worst are platforms for whiners.

406. The Empire State Building: *Soaring splendor.*

It rises 1,455 feet, tower included, and symbolizes New York City's lofty aspirations. This commercial building opened in 1931, then the tallest skyscraper in the world. It soon became a romantic icon, inspiring marriage proposals and figuring in the films *King Kong, An Affair to Remember,* and, recently, *Sleepless in Seattle*. For 40 years the World Trade Center surpassed its heights. Sadly, September 11 put the Empire State Building back on top in Manhattan.

405. Church Suppers: *Thanksgiving in miniature.*

Talk about bounteous buffets: first, pile your plate high with meat loaf, ham, fried chicken, tuna casserole, and pork chops, then work your way through the salads, the vegetables, the fruit and desserts like deep-dish apple pie with cheddar crust. No need to feel guilty: The pastor preached a powerful prayer to set the gluttony in motion, so God forgives you and the millions of other Americans who chow down in houses of worship every week.

404. Medicare: *Rock of aged.* Created in 1965, this is the federal health insurance program for those 65 and older regardless of income or medical history. The program, later extended to younger people with disabilities, helps over 44 million Americans pay hospital bills, medical fees, and drug expenses. In 2008, Medicare benefits totaled $412 billion, accounting for 13 percent of the federal budget. The monumental question now: how to keep this program afloat if trust funds run out in 2017, as some projections foresee?

403. Call-in Radio: *Gab 'n' roll.* Let's hear it for "Tony on the car phone," whose moral outrage against public misdeeds makes for riveting radio during your commute. While there's no rule that you have to be on a cell phone that fades in and out as you drive, it seems that most callers are mobile. Call-in radio prompted the need for the "seven-second delay," a process that allows vigilant radio engineers to delete naughty words and inappropriate ravings before they hit the airwaves.

402. The Yankees: *A synonym for winner.* To say of any American enterprise, "It's like the Yankees," is to say it is the richest and most successful in its field. The New York team has won 27 World Series championships, the most titles of any pro franchise in North America. (The Montreal Canadiens have 24 hockey crowns.) Humorist/publisher Bennett Cerf quipped that rooting for the Yankees is like rooting for U.S. Steel, though we must say that Derek Jeter is more lovable than grouchy robber baron J. P. Morgan was. Joe DiMaggio said, "I'd like to thank the good Lord for making me a Yankee," and millions of American kids wish they could be so lucky.

401. Pooper-Scooper Laws: *A real sky opener.* Freed by these acts of common sense from the need to keep their eyes on the lookout for

Fido's feces, urban walkers can take in the scenery around them. Originating in New York City in 1978, the pooper-scooper movement has spread to other major American (but very few foreign) cities. All you need to clean up after your pet is a little shovel and a Baggie.

400. Tying the Knot: *Safety net for cities.* Cities with high marriage rates are safe; those with low marriage rates are dangerous. Move to San Jose; avoid Detroit.

399. Grand Ole Opry: *Country music's biggest stage.* A live-entertainment sensation first broadcast in 1925, the Opry showcases singing legends like Garth Brooks and current chart-toppers in country, bluegrass, gospel, and more. Every year hundreds of thousands attend its concerts, making the Opry the top attraction in Nashville, Tennessee. Millions more catch it on TV, radio, and the Internet.

398. Cowboy Culture: *Homespun heroism.* It's in our collective DNA: individualism, independence, resilience, and simplicity, the virtues we associate with cowboys and -girls. This rugged life—cattle drives, bedrolls under the stars, chuck wagons—is ingrained in us thanks to countless films and TV  shows, as well as the writings of 19th century novelist Zane Grey and Louis L'Amour in the 20th and 21st. American cowboys learned the ropes from Mexican vaqueros, who showed them how to break broncos, ride herds, throw lariats, and brand cattle. Meanwhile, *American Cowboy* magazine has launched a national effort to have the fourth Saturday in July designated National Cowboy Day.

397. Cable TV: *It's what's on television.* Cable TV dramas like *Burn Notice* and *Entourage* are regularly among the most popular shows on television. Reality programming like *The Real Housewives* of *New Jersey, Jersey Shore, Project Runway,* and *Top Chef* are national phenomena. CNN made news a 24/7 endeavor. ESPN did the same for sports. HGTV taught us how to stage our home to sell it; the Food Network taught us to make Sloppy Joes. Nickelodeon gave our kids mindless entertainment. And the DIY network taught us how to replace our toilets. Cable exploded on the scene in the early 1980s in the United States. Now, about 90 percent of all homes have access to cable stations. Those of us at an age to remember TV dials that gave us access to maybe seven stations still marvel at the choice of more than 250 stations at the touch of a remote. Now that's progress. We think.

396. Buffalo: *Thunder of the prairies.* We almost lost them. Buffalo, or the American bison, once roamed in herds across America's West. Uncontrolled hunting nearly extinguished them. The bison arrived in the Americas by traveling from Asia across land that connected Siberia and Alaska thousands of years ago. You can still catch a glimpse of these burly, shaggy humpbacked creatures that symbolize a bygone era of the American West at Yellowstone National Park in Montana. There, dedicated environmentalists are trying to undo the damage we did. The buffalo population, which once numbered 30 to 60 million, dwindled to about 1,000 before the carnage stopped. Now, about 4,000 bison roam their home in Yellowstone.

395. Buffalo Wings: *Mouth-watering, fire-breathing chicken.* In Buffalo, New York, where this spicy hot appetizer was created in 1964, you can just order wings. Everywhere else, we order Buffalo Wings. They are usually served with ranch dressing and a side of celery sticks to cool the fire they create in your mouth.

394. Creative First Names: *A nation of whatsisnames.* Countries from Belgium to China prohibit parents from giving their children names that aren't on an approved list. In 2009, Germany banned names with double hyphens. In America it seems like every Tom, Dick, and Mary is naming a kid Tomiana, Dickesha, and Marymeeka. Among our favorites—taken from a blog on allnurses.com—are Lufituaeb (*beautiful* spelled backward), Clamidia, Bilirubin, and triplets Shaquita, Tequila, and Markeeta.

393. The Electoral College: *Most interesting unresolved Constitutional conundrum.* This anachronistic system for choosing the President really ought to go, but meantime it has three pluses: it forces candidates to campaign in backwaters like Maine and Idaho, it provides a periodic punching bag for indignant newspaper editors, and it confuses Europeans even more than baseball does.

392. Pies: *Sweet slivers.* The Pilgrims brought us these baked pastries with fillings. Today, the one-crust or two-crust delights have become our most traditional desserts. We consume 186 million pies annually, for $700 million in grocery  sales alone. The American Pie Council reports 13 percent of us favor pumpkin pie; 12 percent, pecan; 10 percent, banana; and 9 percent, cherry. But top choice is apple, at 19 percent, proving the rightness of the expression "as American as apple pie."

391. Money Talks: *The pursuit of happiness in dollars and cents.* In our immigrant society, which values innovation and hard work above caste and class, almost everybody is trying, or at least hoping, to get rich. Even those who fail to hit the jackpot revere rather than revile the winners.

390. iPods: *Where music is moving to.* If you aren't one of those previously mentioned fuddy-duddies who cling to the past, you can now throw out that old stereo and make melted plastic bowls out of your LP collection. More than 250 million iPods have been sold since Apple introduced the tiny way to carry music and videos around in your pocket. Nearly 10 billion songs have been downloaded through Apple's iTunes store. Sure, there are other MP3 players, as these gadgets are generically known, but we salute iPods because they are an American invention and they also got it right best. Apple has also figured out a way to combine the iPod with a cell phone and a computer. It would be the perfect invention if it could wash your laundry too. Oh, well. Enjoy the music.

389. *Sesame Street*: *The idiot box's schoolmarm.* American kids used to start their education with *McGuffey's Reader,* and the shared experienced helped bond the citizenry. Since 1969, toddlers have learned their ABCs by watching *Sesame Street* on public television. Kermit the Frog, Maria, Oscar the Grouch, Bob, Big Bird, Luis & Co. have taught and entertained 80 million kids.

388. *The Wall Street Journal*: *Financial world's bible.* To Managing Editor Barney Kilgore, the journalistic genius who steered this newspaper to greatness beginning in the 1940s, the famed page 1 stories known as leaders had to be understandable to Main Street yet impress bankers and business executives with their insights. They invariably do. Winner of 33 Pulitzers, the venerable paper reaches more than 2 million readers.

387. Mission at San Juan Capistrano: *Strictly for the birds.* Actually, we jest. This adobe mission founded by Father Junipero Serro the year the United States declared independence is a sightseeing treasure. Not only does it preserve the beauty of the 18th century missions along the California coast, but it gives visitors a tranquil respite from the usual tourist traps in

Orange County. It was the seventh mission founded in California by the Franciscans, who eventually established 21 in the territory. In 1845 it was sold to a don for $710, and his family lived there for 20 years. Then, in 1865, President Lincoln returned the mission to the Catholic Church. We might have included San Juan Capistrano even without the annual unexplained flocking of the swallows to rebuild their mud nests in the rafters of the mission church. Studied by scientists, memorialized in song, and shrouded in wonderment, the return of the swallows to the old church in Capistrano takes place on St. Joseph's feast day, March 19. Believers say it is an annual miracle that commemorates St. Francis of Assisi's love of birds and animals.

386. Massachusetts Institute of Technology: *Quiet, geniuses at work.*

People from MIT are assumed to be brainiacs like the alums who founded Intel, McDonnell Douglas, and Hewlett-Packard. And economists like Paul Krugman, who in 2008 was the 73rd grad MIT graduate or teacher to win a Nobel prize. Lulu Liu, class of 2009, spoke for the typical MIT student when she said on her blog, "Armed with a physics degree, I will be in a great position to pursue my ultimate goal in life of just being really good at everything."

385. Coal: *Our energy ace in the hole.*

Coal? The stuff that spews out more carbon dioxide than any other fossil fuel? Yes, coal. Dirty as it is today, we believe our entrepreneurial culture will figure out how to clean it cheaply. Meantime, it is comforting to know the U.S. holds one-fourth the world's recoverable reserves—a 245-year supply at our current production of 1.1 billion tons annually.

384. Instant Celebrity: *The democratization of fame.*

Confirming pop artist Andy Warhol's prediction that everybody will enjoy 15 minutes of fame, this book's authors once held forth on *Good Morning America*. You don't remember us? Then you probably don't remember Harry Whittington, the guy Dick Cheney shot. Or John Mark Karr, the oddball who lied when he confessed to the murder of JonBenet Ramsey. These overnight luminaries can only envy their in-your-face cousins, people famous for being famous, like gossip-column favorite Paris Hilton, the hotel heiress.

383. Savannah: *Dixie charmer.* Settled in 1733, this former cotton town's historic district features lush, parklike squares and moss-draped trees. Tourism has mushroomed in this Georgia town since the 1994 best seller *Midnight in the Garden of Good and Evil*, based on the murder trial of a local millionaire. Visiting the crime scene is de rigueur.

382. Amazon.com: *Earth's biggest store.* What started as an online bookstore has morphed into a place to buy almost anything. Amazon logged more than $14 billion in sales in 2007. It also was positioned as a leader in the future of books, with its handy little electronic "Kindle" that makes it easy to read a book without turning pages.

381. San Diego Zoo: *The cull of the wild.* Always on travel guides' Top 10 lists of zoos around the world and often no. 1, this California zoo has 4,000 animals and 880 species, including one of four panda families in captivity. It pioneered cageless exhibits, giving the animals more space to roam, and visitors clearer consciences.

380. The Filibuster: *The tyrannical minority's defense against the tyranny of the majority.* A single senator can delay passage of a bill indefinitely by speaking on any topic for as long as he or she wishes (or, these days, merely threatening to do so), unless a supermajority of 60 colleagues votes to shut the filibuster down. Huey Long recited shrimp and oyster recipes in a 15½ hour speech in 1935. Strom Thurmond reportedly took a steam bath to dehydrate himself so he wouldn't have to pee while logging the longest filibuster in history, at a futile 24 hours and 18 minutes, against the Civil Rights Act of 1957.

379. *Death of a Salesman*: Pinnacle of American stagecraft. Arthur Miller's 1949 play is still studied and performed worldwide. It's the tragic story of failed salesman Willy Loman, whose distorted view of the American Dream as mere material success shapes his downfall. So far, the Pulitzer Prize–winning drama has been revived three times on Broadway, and recast into a Hollywood film and four television productions.

378. Mary Cassatt's Babies: Maternal impressions. Cassatt (1844-1926), the lone American woman in a circle of French men, stands out as the maternal memory of French Impressionism. Never a mother herself, Cassatt's oeuvre focused on women and children. Her 1893 *The Child's Bath* at the Art Institute of Chicago, and the 1897 *Breakfast in Bed* at the Huntington Library in California, are two of our favorites.

377. Coming-of-Age Celebrations: *Teens in transit.* There is no better reason to party than to celebrate an initiation into adulthood as prescribed by the melting pot's array of ethnic and religious traditions. When youngsters are confirmed in the Catholic Church, they become soldiers of Christ. "Today I am a man" or "Today I am a woman," 13-year-old Jews might proclaim at their bar or bat mitzvah. And the Quinceanera celebrant waltzes her way into Latina womanhood with a court of admirers dancing along. Purists worry that the religious roots of these ceremonies are overshadowed by the partying. We'll drink to that.

376. Reaganomics: *Conservative ying to the New Deal yang.* We need theories from both the left and the right about how to cure an ailing economy because, over the long run, they keep each other honest. FDR is revered by liberals today for his New Deal pump-priming during the Depression that supposedly fixed the disaster visited upon the nation by the hapless Herbert Hoover. Ronald Reagan is esteemed by conservatives for the supply-side economics of lower taxes and less regulation that they claim put an end to "stagflation"—low growth and high inflation—of the Carter years and led to the longest peacetime expansion in American history. Then along came the financial meltdown of 2008, and the liberals were back in the saddle, at least for a while. The great debate continues.

375. "Stopping by Woods on a Snowy Evening": *The Big Sleep.*
On one level this haunting 1923 classic by our most celebrated poet, Robert Frost, is the simple story of a man who pauses on a sleigh ride to admire the beauty of freshly fallen snow. But in the spare, New England style that Frost so ably cultivated, the brief monologue delivers much more: A picture of a traveler through life contemplating his mortality. The memorable last lines: "The woods are lovely, dark and deep / But I have promises to keep, / And miles to go before I sleep, / And miles to go before I sleep."

374. ESPN: *Worldwide leader in sports broadcasting.* It started in 1979 as a wacky idea: Americans were such sport fanatics that they would watch a 24-hour sports cable channel. Who knew? Three decades later ESPN is the most powerful name in sports broadcasting. Headquartered in Bristol, Connecticut, ESPN, Inc., now owned by Disney, has expanded beyond its original sports-reporting channel to include reruns of sporting events on ESPN Classic and all-college programming on ESPNU. It transformed poker and billiards into spectator sports. More importantly, ESPN (coupled with the all-news CNN) gave cable television a huge boost in viewership in the 1980s when the big three networks (CBS, NBC, and ABC) were the only national broadcasters. But thanks to the wackiness of those sports fanatics, it's a whole new ball game.

373. Montana: *Room at the top.* With 6.2 people per square mile, Montana has breathing space and its residents keep alive a Northwest frontier culture that emphasizes independence and outdoor pursuits. Our favorite spot is Columbia Falls, 18 miles south of

Glacier National Park, where the folks are friendly and the weather is almost not too cold.

372. *The New York Times*: Curator of all the news that's fit to print.

The most influential newspaper in the world, the *Times* soldiers on where lesser rivals pull back or close. With its newsroom staff of more than 1,000, it has won more Pulitzer Prizes than any other newspaper and has a better chance than most to weather the Internet's threat. While conservatives bemoan its leftist tilt, readers of all political stripes say they can't do without it in the morning— even though one wag, distressed by its bulk, revised its motto to read, "All the news that fits, we print."

371. The Coast Guard: Sentinels of our shores.

Our smallest armed service with 40,000 men and women, the Coast Guard is entrusted with expansive police powers, often intercepting drug runners or illegal immigrants. The Guard also conducts search and rescue operations, responds to environmental mishaps, and provides navigational assistance. Since September 11, this service has played a vital role in protecting U.S. coasts, ports, and inland waterways.

370. Guggenheim Museum:

Experience the ziggurat. When architect Frank Lloyd Wright designed the Guggenheim's New York museum, he chose a site off Central Park and imagined the winding pyramid that the Babylonians called a "ziggurat." There's always some interesting modern exhibition in the museum on  Fifth Avenue and 88th Street, but you don't go to just see the artwork. You've got to *experience* the building. Take the elevator to the top and then wind your way down through the spiral our kids used to call the giant teacup.

369. Yosemite National Park: *Rugged marvel.*

Opened in 1890, this park covers nearly 1,200 square miles of mountainous terrain in the Sierra Nevada some 200 miles from San Francisco. One of the first wilderness parks in the United States, Yosemite is known for its granite cliffs and waterfalls. Annually, some 3.5 million visitors explore its 800 miles of hiking trails with views of deep valleys, clear streams, giant sequoia groves, and expansive meadows.

368. General Electric: *Storied corporate colossus.*

One of the world's largest companies (it had revenues of $183 billion and profit of $18.1 billion in 2008), GE regularly ranks near or at the top of lists of the best managed, most admired, most recognized, and most worker-friendly companies. It bestrides an industrial empire that ranges from financial services—its core business—to jet engines, health care, computers, and entertainment. It is the only company on the original 1896 Dow Jones Industrial Average that is still there. It took a shaky ride through the recession, but then, it had a lot of fellow travelers.

367. Berkshire Hathaway: *Road to riches.*

Yes, this conglomerate, with holdings from insurance to sweets, faltered in the recession. But for more than four decades it grew by 20 percent a year, providing a windfall for investors. With Warren Buffett, the "Oracle of Omaha," at the helm, it is probably still a good bet. After all, 35,000 people flock to the company's annual meeting, known as the "Woodstock for capitalists."

366. Battling Disease in Africa: *Suffer the poor.*

Mosquito-borne malaria is the leading cause of death in Africa, killing almost 1 million people annually. But this preventable disease can be treated for $10 per person. HIV/AIDS epidemics are spreading through sub-Saharan countries. With just over 12 percent of the world's population, Africa is estimated to have more than 60 percent of the AIDS-infected population. Despite these dire situations, African nations are optimistic that these killers can be controlled. The reason: U.S. humanitarian initiatives begun under President George W. Bush. In 2005 his administration committed $1.2 billion to fight malaria in 11 countries, saving hundreds of thousands of lives. Overall, U.S. aid to Africa has quadrupled from $1.3 billion to more than $5 billion in 2005, and to almost $9 billion for 2010, the largest increase since the Truman administration. Today, the United States and the Bill Gates Foundation are the largest contributors toward fighting HIV/AIDS, malaria, and tuberculosis. No wonder so many Africans love us.

365. Jersey Tomatoes: *Big boys of their class.* Rutgers University launched a "Rediscov-ering the Jersey Tomato" initiative in 2008 be-cause consumers were reminiscing about the way the Jersey tomato *used* to taste. They rein-

troduced the Ramapo tomato, a tasty hybrid. New Jerseyans take their tomatoes seriously. The legislature has been quibbling over whether the tomato can be the state vegetable, given the fact that it is technically a fruit. It's also New Jersey's biggest fresh market crop, totaling $24.9 million annually during the brief harvest season.

364. National Transportation Safety Board: *Reducing the toll.* It investigates all types of transport accidents and recom-mends rules to improve safety. We're bowled over by the board's ability to reconstruct aircraft to learn why they crashed. For exam-ple, after 260 people died in the 2001 plunge of an Airbus 300-600 on Long Island, the agency was able to learn that the design of the rud-der pedal system was a factor.

363. Baseball Caps: *Wardrobe staple for ordinary folks.* Base-ball caps level the fashion playing field, so to speak, shade your face

from the sun and protect it from the rain, and have the added benefit of hiding balding spots, an attribute so treasured by some men that they parade around the workplace in them. Some aficionados decorate them with sequins or patches. Kids often wear them backward, a practice oldsters view as dorky.

362. Great Smoky Mountains National Park: *Ancient elevations.* Covering more than 800 square miles in the southern Appalachians, these are among the oldest peaks in the world, formed 200 to 300 million years ago. For size and climate, this North Carolina–Tennessee borderland is unique in diversity of plants, animals, and invertebrates. Over 10,000 species are documented and another 90,000 may await discovery.

361. Graceful Exits: *Politicians accepting the inevitable.* Popular sovereignty works only if defeated candidates abide by the results. That has been America's tradition since colonial times. Losing incumbents from President to dogcatcher tend to concede graciously and urge their followers to support the winner. "I wish Godspeed to the man who was my former opponent and will be my President," Republican presidential candidate John McCain said in his concession to Barack Obama in 2008. Holding regularly scheduled elections helps the process. Why bother with a coup d'état when you can vote the bums out every few years? Some losing hopefuls do put a sour note in their farewells. A grumpy Richard Nixon told reporters after his defeat in the California governor's race in 1962, "You won't have Nixon to kick around anymore, because, gentlemen, this is my last press conference." He was wrong on both counts.

360. American Robin: *Bird that comes a-bob-bob-bobbin' along.* Spring is coming when you see a robin with its telltale red breast pulling worms out of your still-dead lawn. The state bird of Connecticut, Wisconsin, and Michigan, robins nevertheless make their home in every state.

359. The U.S. Navy: *Anchors aweigh.* Revolutionary hero John Paul Jones told Congress, "I wish to have no connection with any ship that does not sail fast, for I intend to go in harm's way." Today the U.S. Navy projects power for America with 330,000 sailors, 280 vessels, and 3,700 planes. Veteran James Skinner says the skills he learned in a 10-year Navy career—teamwork, discipline, and sense of timeliness—were behind his success as McDonald's chief executive.

358. Temple Emanu-El: *Showcase of the Jewish establishment.* Located on Fifth Avenue in Manhattan, it is the world's largest synagogue, able to seat 2,500 worshippers in its quarter-acre sanctuary under a 103-foot-high ceiling—more than can fill the pews of St. Patrick's Cathedral down the street. The Reform congregation boasts the cream of Greater New York's 2 million Jews, and the Romanesque revival building is an architectural jewel, with massive limestone walls, huge bronze doors that bear symbols of the 12 tribes of Israel, and an arch that wraps around a wheel-shaped window.

357. Folk Music: *Vocal history.* By definition, folk music, which to our ears can embrace bluegrass, country, gospel, jug bands, Appalachian folk tunes, Cajun and Native American songs, is understandable and invites participation. Known as roots music, it has paved the way for later musical developments, including rock 'n' roll, rhythm and blues, and jazz. While folk songs often lament social and political conditions, they also take the form of children's tunes and love ballads. Traditional favorites, like the river shanty "Shenandoah" or the spiritual "Down by the Riverside," have no identifiable composers. But we do know that this music of the working class flourishes in our toughest times. The anthem of the civil rights struggle was "We Shall Overcome." The Vietnam War inspired

Pete Seeger's "Where Have All the Flowers Gone?" To appreciate the richness of this aural heritage, listen to the soundtrack to the 2000 film *O Brother, Where Art Thou?*

356. Amateur Hours: *Reality shows that can make you a star.* If you've got chutzpah, endurance, and a modicum of talent, there are plenty of broadcast opportunities to make you a star. The Fox television hit *American Idol* debuted in 2002. The no. 1 TV program from its inception, *American Idol* weeds through about 100,000 wannabes each year to cull a dozen singers for national exposure. If dancing is your forte, *So You Think You Can Dance* has provided an international platform for would-be Fred Astaires. *Project Runway* features fashion designers; *Top Chef*, *Hell's Kitchen*, and *The Chopping Block* are cooking reality shows, while *America's Next Top Model* targets glamour girls and *The Apprentice* makes celebrities out of business entrepreneurs. Is there no end to this madness? Next thing you know, they'll be making stars out of people who get plastic surgery. Whoops, too late. It's called *Extreme Makeover*!

355. Delmarva Blue Crabs: *Let's hear it for the bays.* For combining good eats with strenuous activity, few things beat an attack on a bucket of crabs from the Chesapeake or Delaware bays. Take a fork, a pick, and a little wooden hammer and do battle with the shells and legs to get the meat. The squeamish might prefer a crab cake. For the best of both, go to Mickey's Crab House in Bethany Beach, Delaware, and wash your meal down with Dogfish Shelter Pale Ale.

354. Ohio: *Microcosm of America.* One of the four most urban states, Ohio ranks third in manufacturing, yet agriculture is the no. 1 industry. Sixty percent of Americans live within 600 miles of the state, making it the crossroads of America (to borrow neighboring

Indiana's hopeful slogan). With its diverse population, Ohio is a swing state in national elections. Tomato juice is its official beverage. Like Virginia, it holds bragging rights as either the first or second most prolific producer of Presidents, having sent either seven or eight to the White House, depending on how you calculate the matter.

353. The Wave: *Stand up and feel the love.* Many claims persist about the origin of the Wave, the most popular cheer in sports history. We'll put our money on professional cheerleader Krazy George Henderson. He swore he got fans to start this rollicking movement—much like synchronized swimming but on land—during a nationally televised playoff game between the Oakland A's and New York Yankees in 1981. Any doubts? Just roll the videotape.

352. The Manhattan Skyline: *New Jersey's most scenic overlook.* The best view of the New York skyline can be glimpsed from the winding entrance ramp that leads motorists through the Lincoln Tunnel into
New York. If you'd rather not subject yourself to that daily traffic mess, there are other vantage points: the Staten Island Ferry, Liberty State Park in New Jersey (where you can view the Statue of Liberty too), or the top of the Empire State Building. By night, head to Brooklyn Bridge Park or the Weehawken waterfront in New Jersey. Bring your camera.

351. Civil War Reenactments: *Blue and gray ghosts.* We abstain from the debate on whether enthusiasts who replay the Battle of Chickamauga and other Civil War confrontations should wear authentic 1860s-style underwear. It's wonderful enough that Americans dressed in battle garb keep the history of the great conflict alive. The first mass outpouring was in 1913, when more than 50,000 Civil War veterans commemorated the Battle of Gettysburg, watched a reenactment of Confederate General George Pickett's uphill charge, and risked being bored to death by the speeches that followed.

350. The U.S. Air Force: *Above all*. That's the motto of the 328,000 members of this high-flying branch of the American military. They serve in the largest, most technologically advanced air force anywhere, with 5,778 manned aircraft, 156 unmanned combat air vehicles, and more than 2,000 ballistic missiles. Originally under the U.S. Army, the Air Force was designated a separate branch in 1947.

349. Make-a-Wish Foundation: *Dream weavers*. The Make-a-Wish Foundation, founded in 1980, fulfills the dreams of deathly ill children. The nonprofit first got into the wish business when it helped a seven-year-old boy dying of leukemia live his fantasy of being a police officer. He died two days later.

348. World Series Moments: *Remembering the boys of autumn*. Since the first one in 1903, the world championship of baseball (yeah, yeah, we know "world" is a misnomer) has produced thrills that fans remember and relate to their children and grandchildren, like tribal stories of brave warriors. New York Yankee Don Larsen's perfect game in 1956. Willy Mays's catch in 1954 for the New York Giants. The two seventh game walk-off home runs—Bill Mazeroski's for the Pirates in 1960 and Joe Carter's for the Blue Jays in 1993. Then there's the flub by the Red Sox first baseman Bill Buckner to end the pivotal sixth game against the New York Mets in 1986. Play-by-play announcer Vin Scully made the call on television, his excitement rising with each phrase. "Little roller up along first.

Behind the bag. Gets through Buckner. Here comes Knight and the Mets win it." (You guessed it. Two of the authors are Met fans. You want a different highlight? Write your own damn book.)

347. Local Boosterism: *The countryside's crack at 15 minutes of fame.* Bellefontaine, Ohio, boasts the first concrete street in America, the shortest street in America, the highest point in the state (a nearby hill), and a "charisma all its own." Though the world may not beat a path to its outdoors, two of us lived there, and all those claims are true.

346. Santa Fe: *Cultural oasis of the Southwest.* Native Americans lived here for centuries. Then the Spanish officially founded the city in 1610, ruling it for 200 years before it became engulfed in our westward expansion. Now New Mexico's capital, its multicultural past permeates its neighborhoods, winding streets, and signature adobe dwellings. Perched at 7,000 feet in the foothills of the southern Rockies, Santa Fe is host to nearly 300 galleries, more than a dozen museums, and a summer opera festival.

345. The Drippings of Jackson Pollock: *A scribbled delight.* The art of Jackson Pollock (1912-1956) could be confused with the penmanship of preschoolers. But make no mistake about this: Pollack's mammoth canvases of paint drippings are worth big bucks. His no. 5 painting sold a few years ago for $140 million, the most ever paid for an American painting. The Wyoming-born artist nicknamed Jack the Dripper was a major force in the Abstract Expressionist art movement after World War II. Pretty much every good modern art collection has its representative splash of Pollock.

344. Bayou Country: *The culture of the wetlands.* A drive along the Bayou Teche scenic byway and other country roads in Louisiana are trips to a place like none other in America. Louisiana has more wetlands than any other state, much of it flooded forests of oak and

cypress teeming with gators. There's Cajun food and folkways, tropical wildlife, and towns that haven't changed much since 1803 when French dictator Napoleon sold the Louisiana Territory to America.

343. Chrysler Building: *Spired inspiration.* Soaring 1,047 feet over midtown Manhattan (earning it the title of world's tallest building for 11 months in 1930 and 1931), it is both an Art Deco masterpiece and a tourist destination. Its famous gargoyles and eagles are modeled after Plymouth hood ornaments and other automobile features, including radiator caps. Its most striking feature from the ground is the 185-foot spire. The Empire State Building is taller, but the Chrysler Building wins the beauty contest.

342. American Ballet Theater: *Pointe well taken.* One of the world's great dance companies, the ABT's repertoire mixes classics *Swan Lake*, for one, with new works like *Push Comes to Shove*. Choreographic genius George Balanchine was a driving force behind this New York group, which was started in 1937. It was Balanchine who nixed restrictive costumes, encouraging dancers to perform in rehearsal-like outfits that accentuated the full dance.

341. Coca-Cola: *The drink that launched a thousand sips.* Actually, trillions upon trillions of sips is more accurate. The 200 countries where Coke is sold consume 1.6 billion servings of Coca-Cola Co. beverages every day, according to the company's Web site. An Atlanta druggist invented Coke in 1885, and yes, it did have extract of coca leaf early on. The recipe for classic Coke is a better kept secret than the formula for the A-bomb. Next to the ideals of the American Revolution, no export has been snapped up with more enthusiasm than Coca Cola. That's not why Coke is one of the best things about America. It's the taste. And those old time six-ounce bottles.

340. Faith in Action: *Believers who put their money where the South is.* Christians, Jews, and adherents of other religions show their mettle when they throw their wallets behind efforts to alleviate poverty in the Southern Hemisphere. In a sweet irony, they recently joined forces with media mogul Ted Turner, who once called Christianity a "religion for losers" but later declared it a "bright spot" in the world.

339. Organ Donation Collaboratives: *Giving the gift that keeps giving.* Nearly 102,000 Americans were awaiting organ transplants as of June 2009. Their prospects were a lot grimmer until recently, when a federal initiative dramatically increased all types of organ transplants. In 2003 the original Organ Donation Breakthrough Collaborative singled out 15 top hospitals that promoted organ donations most effectively, then persuaded the country's 200 biggest hospital trauma centers to copy them. The approach is working: the long-flat 52 percent organ-donor number has shot up to 63 percent. Moreover, the goal is in sight of boosting deceased donor organ transplants to 35,000 annually.

338. Lacrosse: *The sport born of Native Americans.* In this game, 20 kids run around a field swinging long wooden sticks and trying to catch and throw a pretty hard ball. Native American tribes developed the game to prepare their men for war. Today lacrosse is a sport at 1,400 high schools and about 450 colleges.

337. Commemorative Stamps: *Art and history that stick.* The U.S. Post Office Department's first commemorative stamps arrived late. It was the 1893 set honoring the 400th anniversary of Christopher Columbus's landing. Almost 4,000 subjects later, the Postal

Service is commemorating just about everything. A sampling from 2009: Edgar Allen Poe, wedding cakes, early television memories, and undersea kelp forests. You can't beat the price for a work of art: first-class postage.

336. Denali National Park: *Where the wild things are.* The first national park established to conserve wildlife, Alaska's Denali dates from 1917 and is home to 38 species of mammals, including grizzlies, moose, and caribou, plus 160 types of birds from eagles to shrews. A must-see for viewing wildlife in their true habitat, Denali's 6 million rugged acres of subarctic wilderness are dominated by Mount McKinley, part of the Alaska Range and the tallest peak in North America at 20,320 majestic feet.

335. Florida Oranges: *The taste of sunshine.* Ever try to find a juicy Florida orange in Florida? Darn near impossible. That's because most Florida oranges get shipped around the United States or end up as juice. It's a $9.3 billion seasonal industry, October through June. In the dead of winter a plump navel orange is tasty enough to remind you of summer.

334. Winter in Puerto Rico: *Warmth with a Latin beat.* An average high temperature that never dips below 80, a tourist-friendly, swinging, capital city, San Juan, 272 miles of coast and El Yunque—the only tropical rain forest in the U.S. Forest Service's holdings. With cheap direct flights from America's wintry cities, you gotta go.

333. Seattle: *America's most wired city.* We're talking caffeine, not computers: coffeehouses are everywhere. There is so much to see in Seattle: Puget Sound and Lake Washington and the Chittenden Locks that connect them, the salmon runs, the flotillas of

houseboats on Lake Union, the Pike Place Market where fishmongers hurl fish through the air, Pioneer Square, the Space Needle, and Mount Rainier (on a clear day). If you get bored, the Cascade Mountains lie to the east, the Olympic Mountains to the west, and the San Juan Islands are just a ferry ride away. Canada and Alaska beckon.

332. Local Festivals: *Happy hoopla*. Americans celebrate almost anything. Take Ketchikan, Alabama's timber carnival; the Bourne, Massachusetts, proud scallop fest; and the Cobden, Illinois, paean to peaches. Conventional festivals, those, but there are some head-scratchers like the arts festival in Black Rock, Nevada, a town that exists for this gathering just one week a year, or the Pandemonius Potted Pork Festival that keeps Austin, Texas, in stitches—and Spam—around April Fool's Day.

331. Convenience Stores: *There when you need 'em*. Fifty billion customers annually shop at convenience stores. The beauty of shops like 7-Eleven, WaWa, Piggly Wiggly, Circle K, Quick-Chek, Cumberland Farms, and others is that you are in and out in under four minutes, according to National Association of Convenience Stores statistics. They annually sell $187 billion in tobacco products, $18 billion in lottery tickets, $20 billion in beer, and $1.2 billion in potato chips. You pay a premium, but when it's the only place open on Thanksgiving, you can find that can of cream of mushroom soup to complete your string bean casserole.

330. Intracoastal Waterway: *Sheltered passage*. Boats can travel on this 3,000-mile series of natural and man-made water-

ways inland from the Atlantic and Gulf coastlines from Boston to Key West, from Apalachee Bay, Florida, to Brownsville, Texas. For the queasy, it beats braving the rough ocean. The Army Corps of Engineers tries to keep the channel 12 feet deep.

329. Little League: *Pint-sized great American pastime.*

It began in 1939 in Williamsport, Pennsylvania, with one league of 30 players. Today this baseball organization has more than 7,000 leagues worldwide and 2.2 million players (softball leagues reach another 2.6 million). It's a great way for boys and girls ages five and up to learn the basics of baseball plus discipline, teamwork, and physical fitness. Some doting parents love it too, and suffer withdrawal at season's end.

328. Apple Inc.: *Innovative to the core.*

What started as a homemade computer company grew into a $50 billion electronics giant. Apple's Macintosh computer, which debuted in 1984, brought graphic design capabilities to our desktops. The Mac also transformed words like "mouse" and "icon" into computer terms. Macs are usually thought of as the *other* computer, compared to the more popular Microsoft-driven PCs, but Apple aficionados would rather fight than switch. Even vocal Mac haters were converted to Apple lovers when Apple introduced a revolutionary digital music player, the iPod, in 2001. Then Apple took a bite out of the cell phone market, introducing the iPhone in 2007. Its 2010 innovation is the portable iPad mobile computer. Next on the horizon: iTV.

327. Hospice Care: *Death with dignity.*

The idea that comfort and pain management is the way to treat people when death is near flowered in England and has been embraced by America, which has 4,700 hospice programs. At any given time almost 1 million Medicare subscribers take advantage of this kind of care. A cancer patient wrote on the Hospice Foundation of America's Web site that he told his doctor that when the time came, "I want him to release me to the hospice so that I can die in peace and dignity. I don't fear death, I fear dying."

326. Covered Bridges:

Romantic relics of horse-and-buggy days. Why were they covered? Not to keep snow off the floor, but to protect the supporting timbers against moisture and rot. Spread across rural America, they take their most elegant shapes in Vermont, hugging the rocky streams and rivers that cascade below them and the maples and beeches and pines that stand guard over them. If you want to view two of these icons in an hour or so, visit Waitsfield, Vermont, home to the landmark Great Eddy Bridge, built in 1833, and the smaller but gorgeous Pine Brook Bridge, opened in 1855.

325. The New York Public Library: *Lionizing learning.* Amazing that by the end of the 19th century, celebrated New York City still had no facility that truly could be called a public library. Donors soon corrected that, funding what would become the only library we know with world-acclaimed research centers and 87 neighborhood branches, all free to the public. Its majestic Fifth Avenue Beaux Arts center, guarded by impressive stone lions, houses copies of the Gutenberg Bible and Jefferson's Declaration of Independence.

324. American English: *A likable lingo.* Colour, honour, favourite: who needs all those u's anyway? In Britain, if you knock someone up, you rap on their door. There, a boot is the trunk of a car; an aubergine is an eggplant; bangers are sausages; bangs are a dirty way of saying a couple rolls in the hay; and fringe is bangs. But now 310

million Americans account for two-thirds of the English-language speakers worldwide. That majority, coupled with the prominence of American television, Hollywood movies, and an American presence on the Internet, are steamrollering the British way of speaking. For the record, the term "stiff upper lip," used to describe the stoicism of our friends across the Atlantic, was coined in the USA.

323. Big City Subways: *Weaning us off cars.* We use the term

"subway" as shorthand for big city rail lines that can be above-ground too. The largest, oldest, and dirtiest, New York's 229-mile system, takes riders on 1.6 billion trips a year. The next busiest is the Washington Metro, followed by systems in Chicago, Boston, Philadelphia, San Francisco, Atlanta, and Los Angeles. The safety records of all these services are good, though New York had a problem in 1933 when King Kong ripped an elevated train off the track.

322. Rednecks: *True blue Americans.* These mostly Scots-Irish

country people have endowed our culture with more than country music, Christian fundamentalism, and a love of guns and NASCAR racing. They have also given us our rugged individualism and our fighting swagger. By one expansive reckoning—people of Scots-Irish ancestry who grew up poor in small-town and rural America—famous rednecks include frontiersman Daniel Boone and Presidents Andrew Jackson, Ronald Reagan, and Bill Clinton. The origin of the term "redneck" is uncertain; some say it hearkens back to a movement in Scotland whose acolytes wore red scarves, while others maintain that it refers to the sunburned necks of farmers. Senator James Webb of Virginia wrote a book about them in which he said, "We face the world on our feet and not on our knees. We were born fighting."

321. Chili: *Some like it hot.* This piquant stew

of beef, tomatoes, and beans originated not in Mexico, but San Antonio, Texas. There, beginning in 1840, cooks blended dried beef, beef fat, chile powder, spices, and salt, then pressed this into a brick to be reconstituted with water—perfect takeout vfor California-bound gold prospectors. Today, southwesterners chuck in stew beef, Californians substitute chicken and add wine, and Cincinnatians plop this American classic on spaghetti.

320. Bumper Stickers: *Your car makes a statement.* In Europe, bumper stickers identify the vehicle's country of origin, but here in the United States we want the entire highway to know exactly what we think. Our favorites: BEER IS NOW CHEAPER THAN GAS. DRINK! DON'T DRIVE; GOD IS COMING AND SHE'S REALLY PISSED; IF YOU BELIEVE IN TELEPATHY, THINK ABOUT HONKING, and MY MOTHER WAS A MOONSHINER AND I LOVE HER STILL! Honorable mention goes to the old chestnut, MY ROTTEN KID BEAT UP YOUR HONOR STUDENT!, which still cracks us up as we fly by it on the road.

319. The Colorado River: *White water wonder.* Its precipitous drop of 9,000 feet over its length makes for some of the best rapid rafting in the country. But drought has made the management of America's seventh longest river a test of how the country will preserve its precious water. Before the Colorado is gone, go see it trickle through the Grand Canyon.

318. Presidential Oratory: *Intersections of hope and history.* The most memorable lines conjure up visions of a better America, even a better world. "Steer clear of permanent alliances," Washington pleaded on leaving office. "This nation, under God, shall have a new birth of freedom," Lincoln re-  solved at Gettysburg. "The only thing we have to fear is fear itself," FDR reassured an uncertain nation. "Ask not what your country can do for you," JFK beseeched Americans at the height of the Cold War. "Mr. Gorbachev, tear down this wall!" Reagan demanded toward the end of it. Stirring words that stir us still. (We are confident Obama will produce lines to rival theirs sometime in his presidency.)

317. Silent Spring: *Literary launch of modern environmentalism.* A staple of high school reading lists even today, *Silent Spring* exposed the damage that chemicals can wreak on our ecosystem. In her 1962 book, marine biologist Rachel Carson described how the powerful pesticide DDT entered the food chain and built up in fatty tissues of animals and humans, causing cancer and genetic damage. Chemical companies derided her, but a presidential panel vindicated her, and DDT became history. A legacy of her book: creation of the Environmental Protection Agency.

316. Upward Bound: *Another rung on the ladder out of poverty.* A remnant of President Lyndon Johnson's War on Poverty, the Upward Bound program prepares educationally disadvantaged, impoverished students for higher education. Carved out of the Economic Opportunity Act of 1964, Upward Bound sends about 92 percent of participants to postsecondary education. It serves about 56,000 people annually, preparing them for their SATs, providing instruction and enrichment, and convincing them that a college diploma will help them rise above their circumstances.

315. Air-Conditioning: *Making summer less gritty in the city.* And in suburbs, deserts, cars, and factories too. America generates almost 4 trillion kilowatt hours of electricity each year, as much as Russia, Japan, France, Germany, India, and Canada combined. Much of that energy goes for a good cause: keeping us cool. Yes, we know our addiction makes us less hardy than we should be. But without air-conditioning, population shifts to hot states like Arizona and Florida would never have occurred. Life is too short to sweat the small stuff or suffer in a pool of perspiration in the beds of August and pray in vain for a breeze.

314. Biggest Transfer of Wealth in History: *Generation X's cash cushion.* We've seen a lot of estimates, but the most bruited about is $41 trillion. However you cut it, the sons and daughters of the we-want-it-all Baby Boomers apparently will get it all as their moms and dads go to their heavenly reward and bequeath them a big, fat earthly one.

313. The Salvation Army: *Salvaging hope.* Every Christmas-time its workers man street corners and ring bells, blessing donors who drop money in their red kettles. This fund-raising started in 1891 in San Francisco, with a sign reading KEEP THE POT BOILING, then spread throughout the world. Today this peaceful army raises millions of dollars to provide food, clothing, medical care, and toys for more than 2.9 million Americans in need. Moreover, its kettles allow the evangelical organization to battle addiction, poverty, and homelessness all year long.

312. Personal Space: *We don't get too close!* Anthropologists say we Americans need at least 18 inches of personal space. Citizens of Mediterranean and Latin  American countries get too close for our comfort. Our love of personal space is really quite practical. That 18-inch separation is enough distance to keep spittle from flying on your face in the heat of conversation. Although we are getting better at hugging and touching, Americans eschew the social air-kissing (two in France, three in Belgium) Europeans love.

311. 4H Clubs: *The udder side of teenage life.* The founders in the early 20th century had a brilliant idea: teach scientific methods to farm children who would then pass the info to their parents. The familiar 4H abbreviation stands for the head, heart, hands, and health each member pledges to society. Today the movement has moved from the farm to the suburbs, and occasionally to inner cities like Brooklyn.

310. Hoover Dam: *Concrete wonder.* "Gee, this is magnificent," Franklin D. Roosevelt said on seeing the brainchild of Herbert Hoover that straddles the Colorado River at the Arizona-Nevada border. When completed in 1936, it was the world's largest electricity generator and biggest concrete structure, 726 feet high. Though a relative pygmy today, it attracts 1 million visitors a year because of its historic significance and architectural gracefulness.

309. Charleston: *Southern jewel.* Britain established this, its first colony south of Virginia, in 1670, and present-day Charleston shows its pedigree. Older sister to Savannah, Georgia, Charleston's architecture is delightful, its gardens inviting, its wrought-iron gates works of art. Then there's Fort Sumter in the harbor, where Confederate troops attacked Union forces, igniting the Civil War.

308. The Works of Eugene O'Neill: *The genius cometh.* The only American playwright to win the Nobel prize wrote about the tragedy of being human, and his often revived classics—*The Iceman Cometh* and *Long Day's Journey Into Night*—burn with passion and desperation almost six decades after his death in 1953. Visitors to his home near Danville, California, can see the room where he penciled six masterpieces with a trembling hand.

307. Sitcoms: *The stuff that makes the boob tube the boob tube.* Most situation comedies that have come and gone since regular television programming began in 1950 are eminently forgettable. Still, the formulaic 30-minute contained episode format has endured since *The Burns and Allen Show* established sitcoms as the backbone of the prime time line-up. We pay homage here to some of the great ones, from *I Love Lucy*, the madcap 1950s comedy showcasing Lucille Ball, to *Seinfeld*, a show about nothing that dominated television in the 1990s, to today's clever *The Office*, featuring Steve Carrell as a bumbling paper company manager. American sitcoms export our culture around the world. They live on and on in syndication and in perpetual reruns on YouTube.

306. Rooting for the Underdog: *Rudy, Rudy, Rudy.* Our immigrant ancestors fought to succeed, and the transmission of their spirit explains why we cheer for any team that is playing the Yankees or whatever upstart software company is challenging Microsoft. It's why we flock to movies like *Rocky, Slumdog Millionaire, Norma Rae,* and *Rudy,* the true story of the skinny kid who spent four years on Notre Dame's football practice squad and finally got into a game.

305. Science Fiction: *Journeys to weird places.* Storytelling that pushes us to the limits of our imagination with themes like space and time travel, extraterrestrials and fantasy worlds, can't help but attract a huge following. True, its modern form has roots in Europe in the likes of Jules Verne, the French author of such classics as *A Jour-*  *ney to the Center of the Earth,* published in 1864, and Arthur C. Clarke, the British author whose literary masterpieces included *Childhood's End* and *2001: A Space Odyssey* in the second half of the 20th century, as well as in the U.S., as exemplified by the prolific Isaac Asimov, who turned out the famous *Foundation, Galactic Empire,* and *Robot* series. Even so, Americans evangelize the cult best with blockbusters like the *Star Trek* franchise and the *Star Wars* and *Lord of the Ring* film trilogies.

304. Town Meetings: *Democracy 101.* How many municipal meetings have we sat through where political blowhards debate issues then vote along party lines? Oh, give us the bracing wind of New England town meetings, with their packed auditoriums, spirited Q&As, and the voters—not elected officials—determining the community's actions.

303. Craigslist: *The good idea that scared the wits out of classified ad managers.* Want a free turtle in Clifton, New Jersey? How about a ride to Vermont? Want to sell your house? Check out the Internet's free alternative to classified ads. Founded in 1995 by Craig Newmark of San Francisco, it created a stir among ethicists when desperate children used it to solicit for a much-needed kidney for their father.

302. Le Bernardin: *Best restaurant in the country.* Look up the ratings yourselves; this New York City eatery has earned so many "no. 1" and "four-star" accolades it is starting to seem a little ho-hum. Besides, gourmets sampling dishes like shrimp poached in truffle foam and Kumamoto oysters with green apples can arrive at the same conclusion through independent research. A couple can easily spend $1,000 for a tasting menu with wine pairings plus after-dinner drinks and a tip. For just $1 million they could dine there every night for two and a half years.

301. Superman: *Favorite fantasy figure.* Be honest. If you could fly, see through walls, and be impervious to bullets, would you slave at a newspaper and date Lois Lane? Sure she's got a nice figure, but as a man of steel you could have the most beautiful babes flocking to your, uh, door. You could hit 800-foot home runs for the Yankees, find gold two miles deep, and be richer than zillionaire Warren Buffet. Too bad the President gig would be out, since *60 Minutes* would expose you as an illegal alien named Kal-El who was airmailed to Kansas from Krypton in the 1930s. You could, however, be governor of California, like that other strong foreign guy.

300. DNA Testing: *Elementary, my dear Watson.* The discovery of genetic building blocks by James Watson and Francis Crick in 1953 makes it possible to know for sure whether your daddy is your daddy. And for crime scene investigators to find suspects. And for lawyers to prove their clients were framed. In 2009 the Innocence Project associated with Cardozo Law School in New York said DNA testing had led to the exoneration of 241 convicts. Now the bad news: savvy criminals are spreading other peoples' DNA at crime scenes to confuse the cops. Sounds to us like a CSI episode.

299. Local Zoning Laws: *I wanna be a good neighbor.* Zoning laws keep porno shops and chicken coops out of residential areas and condos out of manufacturing districts. They don't strip people of their right to strip off their clothes, in other words, they just restrict the places where they exercise that right.

298. Amtrak: *Harbinger of a new rail age.* Spurred by higher gas prices, the hassles of air travel, and the joy of hushed rides though woods and fields, far from traffic jams and Golden Arches, Americans are using our intercity train service in record numbers (28.7 million passengers in 2008). Though Amtrak serves 400 destinations in 46 states, it is of pitiable size compared with train systems in most industrialized countries. That is good news, because there is so much room for growth.

297. Hershey Bars: *Meltingly delicious.* During World War II, GIs doled out these milk-chocolate calling cards to war-weary Europeans. Milton Hershey developed the recipe in 1900 to let average folks get a taste of what the rich took for granted. He built the world's largest chocolate factory, together with an entire community, in what's now Hershey, Pennsylvania.

296. Adult Education: *A commitment to lifelong learning.* On any given weeknight in classrooms around the country, grown-ups go back to school. They take life enhancement courses in foreign languages, computer technology, exercise, arts and crafts, and job skills. Adult education has also allowed 15 million people to earn a General Equivalency Diploma (GED) instead of a traditional high school certificate.

295. The John Wayne Model: *Men doin' what they gotta do.*
Thirty years after movie star John Wayne's death, millions of Americans cling to the notion that he personified true manhood. Not John Wayne the person—a sexist, Red-baiting, careerist who dressed up in cowboy costumes and opted out of military service in World War II. We're talking about the character he played in most of his 170 movies: a big man of courage, who spoke low, spoke slow, and would be violent in defense of good if pushed around. Like J. B. Books in *The Shootist*, who said, "I won't be wronged, I won't be insulted, and I won't be laid a hand on. I don't do these things to other people and I require the same from them." We'll saddle up with that creed any day, pilgrim.

294. Our Love of Superlatives: *The* absolutely most *embedded trait of our* incredibly unique *national character.* Americans boast variously about having the world's tallest tree (at 367½ feet), crookedest street (Snake Valley in Burlington, Iowa), hugest meteorite crater (4,150 feet across and 150 feet deep in Winslow, Arizona), biggest tricycle, smallest church, most dazzling canyon, and deepest or biggest certain-categories-of-lake. But we take our licks gracefully, ceding bragging rights for most towering skyscraper to upstarts over the years and for the hottest temperature—134 degrees Fahrenheit in Death Valley's Furnace Creek in 1913—to Al 'Aziziyah, Libya, which stole the crown with 136 degrees in 1922.

293. States' Rights: *What's wrong with that?* A lot—if that noble principle of federalism is perverted by prejudice. The Jim Crow South invoked it to defend segregation, giving the 10th

Amendment a bad name. These days, though, we're glad that politicians on the left and right are reviving that defense against encroachment of our rights after six decades of Washington-knows-best mentality. In fact, in progressive causes like restricting greenhouse gases and cracking down on predatory lenders, the states are often way out front of the federal government.

292. Hudson River: *Mighty mover.* The Hudson flows some 300 miles from the Adirondacks down to New York and the Atlantic. Henry Hudson discovered it in 1609 while seeking China. As the American colonies grew in the 17th and 18th centuries, it became our first great waterway, a river thoroughfare for all manner of goods and people. Today, tugboats, ferries, even cruise liners sail it past the grandeur of Manhattan. It was also the scene of one of the most remarkable (some say miraculous) feats in American aviation history: the flawless landing of a 50-ton jetliner by the quietly heroic pilot, Captain Chesley B. Sullenberger III, in January 2009.

291. Our Memory of the World Trade Center: *Rest in peace.* As a people we have a long collective memory. We remember the Alamo, the *Maine*, Pearl Harbor, the Kennedy Assassination, the *Challenger* explosion. Today's Americans will always remember where they were when the towers fell on September 11, 2001.

290. Mount McKinley: *Or is it Denali?* McKinley is the name in the geography books, though Alaskans and climbers prefer the Native American name Denali, which means "the high one," apt for North America's tallest mountain, 20,320 feet high. An expedition in 1913 was the first to scale the higher of its two peaks, and climbing it is a macho rite of passage for men and women adventurers. More than 100 people have died trying to reach the top.

289. Laser Technology: *A thousand beams of light.* Developed—where else?—in American laboratories, laser devices have endless applications that make our lives easier and safer, from scanning bar codes and storing information in compact discs to performing surgery and improving the accuracy of our weapons systems.

288. The Newport Jazz Festival: *Rhythm and riffs.* Jazz history-in-the-making has captivated crowds in this Rhode Island resort town nearly every summer since 1954. When 11,000 attended the first two-day event, backers knew they had a winner. Since then, live performances recorded by the likes of Ella Fitzgerald, Billie Holiday, Carmen McRae, Muddy Waters, Dave Brubeck, and Ravi Coltrane have been heard by millions.

287. Rugged Individualism: *Going it alone.* This is mostly a guy thing involving hiking boots and isolation. Americans reserve the right to march to a different drummer and to blaze a solitary path, both physically and philosophically. British poet John Donne exclaimed, "No man is an island," but we prefer the wisdom of our beloved Robert Frost, who celebrated taking the road less traveled "and that has made all the difference."

286. Independent School Boards: *Keeping education local (though sometimes loco)*. Except for a few misguided souls who try to ban books or steer contracts to their pals, the 95,000 members of American school boards—usually unpaid—serve selflessly to give their communities the finest education taxes can buy. The best boards have a tradition of remaining aloof from local government politicians. The worst part of the job: fending off calls from parents whose kids didn't make a team and want the coach fired.

285. Grizzly Bears: *Monarchs of the wild.* Once headed for extinction in the lower 48 states, these magnificent beasts have rebounded to nearly 1,200 (a fraction of Alaska's population of 30,000). Magnificent as they are, it is best to revere them from afar (as California does by adorning its flag with one). If you encounter *Ursus arctos horribilis* in the wilderness, back away and sing off-key at the top of your lungs. Don't flee: males weigh up to 800 pounds and sprint at 35 mph.

284. St. Patrick's Day: *Green rules.*

About 34.5 million U.S. residents claim Irish ancestry, almost nine times Ireland's 4.1 million population. That's why St. Patrick's Day, March 17, is a big blowout here and lots of non-Irish join the celebration. The event, a holy day in Ireland, honors St. Patrick, who introduced Christianity there in the fifth century. From the Emerald Isle we get the color of the day. People dress in green and quaff green-dyed beer. They eat traditional corned beef and cabbage. Massachusetts

has the highest concentration of Irish-Americans, so Boston's St. Patrick's Day Parade is an institution. But nothing equals the New York version, which boasts about 2 million revelers and more than 150,000 marchers with bands and bagpipe players from across the country. The Fifth Avenue procession begins at 44th Street, winds past St. Patrick's Cathedral at 50th Street, then ends at 86th Street.

283. The Beach Boys' Good Vibrations: *They're still giving us ex-ci-ta-tions.*

They are now a wrinkled bunch of balding surfers, but in their heyday—circa 1965—the Beach Boys was the no. 1 American band. Their harmonic sound celebrated pretty girls, sandy beaches, and endless summers. The quintet, led by the musical genius yet perpetually troubled Brian Wilson, garnered 36 top 40 hits and is the best selling American band ever. While "Good Vibrations" is their signature song, we prefer the sweet sound of 1968's hit, "Do It Again."

282. Great Salt Lake: *Watery desert.*

The largest lake west of the Mississippi River, 75 miles long and 35 miles wide, has no outlet. Streams with small amounts of salt feed it and the water escapes by evaporation, leaving the salt behind. We recommend kayaking to Antelope Island to see the 500 bison, the country's largest publicly owned herd.

281. Central Park: *Magic in the midst of Manhattan.*

To stroll into the 853-acre oasis of green is to slip away from an urban tempest into an enchanted garden. Whether gawking tourists or worldly locals, the 25 million visitors to the park each year feast on its visual pageantry, from the lakes with rowboats and the woodlands like the Ramble, to Belvedere Castle and the 3,500-year-old Egyptian obelisk behind the Metropolitan Museum.

280. The Hubble Space Telescope: *Astral sleuth.* Launched by NASA in 1990 and named after pioneer astronomer Edwin Hubble, the telescope is an observatory the size of a large school bus, orbiting above earth's atmosphere. It has produced extraordinary images of space and revolutionized astronomy by solving great mysteries, like the age of the universe—13.7 billion years. Thanks to five Space Shuttle missions, astronauts' repairs and installation of new instruments have dramatically extended Hubble's powers.

279. Microbreweries: *This suds for you.* Every town used to have its own brewery, and maybe those days are returning. The Brewers Association says Americans guzzle 1.7 million barrels of beer bubbling out of 1,300 microbreweries and brew pubs each year. These beers are the antidote to the boring output of Anheuser-Busch and its ilk. We salute the imagination and skills of producers like the Hoppin' Frog Brewery in Akron, Ohio, maker of Silk Porter, Bodacious Oatmeal Russian Imperial Stout, Smashing Berry Ale, and our favorite: Hoppin' to Heaven IPA.

278. Idaho Potato: *This spud's for you.* We Americans chow down about 140 pounds of taters a year. One quarter of the nation's potato crop comes from Idaho. There's something about the northwestern climate and volcanic soil that is conducive to making tubers grow. About 60 percent of the $710 million annual crop is chopped and frozen into french fries and hash browns. We like them best baked and slathered in butter.

277. Golf Courses: *Splendor in the grass.* Mark Twain joked that golf was a good walk spoiled. Tell that to the players who walk the

Augusta National course in Augusta, Georgia, every year for $7.5 million in prize money. For the rest of us, down to the lowest duffer happy to score 120, time spent on one of the country's thousands of public and private courses is reward enough. We leave the hustle of the streets. All is quiet but the thwack of a drive, though an occasional curse of frustration is allowed. These are heaven's annexes, from subway-accessible Van Cortland in the Bronx, the first public course in America, opened in 1895, to Princeville at Hanalei in Kauai, where errant shots sail into the Pacific, 300 feet below the cliffs.

276. Bargain Prices Everywhere: *Lifeline to the poor, bonanza for cheapskates.* The United States is a shoppers' paradise. Just ask visiting Europeans who marvel at the price of everything from gasoline and used cars, to groceries and dinners out, to off-label booze and home appliances, to 100 acres of land in the Adirondacks. Some fly to New York just to stuff their empty suitcases with designer clothes they buy at discount stores and use the savings to pay for their trip.

275. The White House: *A home to vie for.* Symbol of the American presidency, the white colonnaded mansion at 1600 Pennsylvania Avenue is recognizable the world over. It is the only head of state's home open to the public, and free at that. Completed in 1800 of Aquia sandstone in late Georgian style, it has housed every U.S. President since John Adams. Battered by political storms, the edifice also has survived a British torching in 1814, a West Wing fire in 1929, and Bill Clinton's 800-plus political donors treated to overnights in the Lincoln bedroom.

274. Summer Camps: *Learning to get along.* "Healthy relationships grow, well-being blooms, and nature beckons." So says the 7,000-member American Camp Association. Want your kids to learn water sports, be better Christians, take cello lessons, or get away from the city's mean streets, compliments of the Fresh Air Fund? Send them to camp. It's the place where children learn independence in bucolic surroundings. Woody Allen joked that his parents sent him to an interfaith camp, where he "was savagely beaten by children of all races and creeds." Today they would send him to comedy camp.

273. The Second Green Revolution: *Famine-buster part II.* Part I was sparked by the late Norman Borlaug, who developed strains that tripled grain production in the second half of the 20th century. The second part is now unfolding, with American agronomists at the center of the movement to employ a mix of policies including Borlaug's use of hardier seeds, the development of genetically modified crops and water-saving technologies, organic farming, and improved food-storage techniques to feed a world population that is expected to climb to 9.7 billion in 2050 from 6 billion in 2000.

272. Ballroom Dancing: *Waltz this way.* Public television has offered synchronized dancing competitions for almost three decades, but the popular ABC reality show, *Dancing with the Stars*, has kicked this artful exercise up several notches. Long considered an old folks' hobby, ballroom dancing now is inspiring kids to take up the foxtrot, waltz, swing, cha-cha, rumba, and tango. More than 300 colleges provide these dance programs, with 1,000 more planned.

271. Liberty Bell: *Our flawed symbol.* It  wasn't all it was cracked up to be. The Liberty Bell—forged in 1752 to commemorate the 50th anniversary of the Pennsylvania Constitution—was supposed to proclaim liberty. But every time it was rung, fissures in the metal deepened. Philadelphia authorities kept trying to repair it until 1856, when they gave up and declared it unringable. No one seems to mind the flaw: about 1.5 million people annually visit this cracked but much loved 2,080-pound clanger of copper, tin, and lead, with traces of zinc, iron, silver, antimony, arsenic, gold, and nickel. It can be found across from Independence Hall, at Sixth and Market streets, in the heart of Philadelphia's historic district.

270. That First Electric Guitar: *The instrument of memory.* They can be Gibsons or Fenders or Kmart specials. And a huge percentage of them are stashed in closets after the first month, never to be played again. Yet for musicians who never stop playing—even those who appear to the world to be accountants, laborers, or plumbers—the first guitar is like a first love affair: the memory gets sweeter with age.

269. The Ohio River: *Waterway of childhood dreams.* Largest tributary of the Mississippi, early gateway to the West, border between free and slave states (whence the phrase "sold down the river"), the Ohio was called by President Thomas Jefferson "the most beautiful river on earth." One of us grew up along its Indiana shores, swam and rafted its waters, and visited a local writer/artist in his cabin on its Kentucky banks. The artist, Harlan Hubbard, who died in 1988, described the Ohio he portrayed in one of his paintings as "a golden current coming forward from the distance, the landscape a patchwork of sun and shadow."

268. The United Negro College Fund: *Narrowing the education gap.* This organization raises tuition money for more than 60,000 black students a year, 60 percent of them the first in their families to attend college. It also provides operating funds for 39 private, historically black colleges and universities. Founded in 1944 in Fairfax County, Virginia, its oft-quoted motto is, "A mind is a terrible thing to waste."

267. MOMA: *Where the wacky, weird, and avant garde hang out.* New York City's Museum of Modern Art is the fourth best museum in the United States. Behind the A-Team of the Metropolitan, the Getty, and the National Gallery, MOMA showcases the eccentricities of modern art. You can experience Dada, soak yourself in surrealism, or debate the value of monochromism. Vincent Van Gogh's *Starry Night* is its most famous painting, but we like Marc Chagall's *I and the Village* or Piet Mondrian's *Broadway Boogie Woogie*. Stop by on a Friday evening when admission is free.

266. National Oceanic and Atmospheric Administration: *Come rain or come shine.*  Mark Twain complained that "everybody talks about the weather, but nobody does anything about it." This agency begs to differ. Using the latest research and instrumentation, its scientists deliver daily weather forecasts and severe storm warnings as well as monitor the climate, all to protect lives, property, and natural resources. In fact, theorists figure the group's work has an impact on more than one-third of America's gross domestic product. The agency dates from 1807, has offices in every state, and is a world leader in scientific and environmental studies.

265. Greenbacks: *The buck never stops.* You want money with fancy colors and pictures of obscure people? Go abroad. Like to Albania, where the rainbow-colored 200 lek note bears a portrait of poet Naim Frasheri. Though some red has crept in, U.S. currency has remained predominantly a comforting green since 1861, with head shots of our great political leaders. The obscure wild card is Lincoln's Treasury Secretary, Salmon P. Chase, who founded the federal currency system and is on the $10,000 bill. U.S. cash printed on 100 percent rag paper is so durable it survives repeated washing in pants pockets, though the government frowns on other forms of money laundering. If your bucks tear or get crunched, don't worry. The U.S. Treasury redeems $30 million worth of mutilated currency annually.

264. The Alamo: *Symbol of heroic struggle against insurmountable odds.* This San Antonio mission witnessed the ultimate sacrifice of nearly 200 fighters, including Jim Bowie and Davy Crockett, for Texas's freedom against Mexican rule. They lasted 13 days before General Santa Anna's army triumphed on March 7, 1836. The Alamo is the state's premier tourist attraction, drawing more than 2 million people each year.

263. Mall of America: *If you build it, they will shop.* Put your walking shoes on for this mall-turned-tourist attraction that skeptics scoffed at as folly when it opened in 1992. You have to trek 4.3 miles to window shop at all 520 stores. Located about 15 minutes from downtown Minneapolis–St. Paul, it ranks as the no. 1 tourist site in the Midwest, attracting 40 million visitors annually.

262. Susan G. Komen for the Cure: *The pink ribbon charity.* The organization named for a Texas woman who died of breast cancer in 1980 raises more than $250 million in a typical year—the most of any breast cancer charity. The chance of a U.S. woman developing the invasive form of the disease at some time in her life is about one in eight, but survival rates have grown thanks to awareness programs of organizations like Susan Komen's and the American Cancer Society.

261. African-American Literature: *The accent is on literature.* From Harriet Wilson's antebellum novel about slavery to Diane

McKinney-Whetstone's chronicles of our times, African-American writing has enriched American culture and served as points of light for understanding the black experience. Toni Morrison lamented, "Black literature is taught as sociology, as tolerance, not as a serious art form," which it is in the hands of Langston Hughes, Richard Wright, Ralph Ellison, James Baldwin, Amiri Baraka, Alice Walker, and Morrison, one of only nine U.S.-born writers to win the Nobel prize for literature.

260. Art of Winslow Homer: *True to Nature*.
Paintings by this master colorist (1836-1910) capture the foam-flecked, often perilous seas off the Maine coast; the exuberant greens, reds, and blues of the steamy Tropics; and his frequent man-in-nature theme of hunters, fishermen, and their quarry. Our favorites: *Fox Hunt*, his largest oil, showing crows attacking a fox in the stark snow, and *Life Line*, the dramatic rescue of a nearly drowned woman.

259. Nebraska: *Model of fiscal restraint*.
The state treasurer boasted about his government's stewardship of taxpayer money in a *Wall Street Journal* op-ed article when the latest recession was going full throttle. The case he made was pretty impressive: a foreclosure rate of one in 25,500 homes versus a national rate of one in 500; an actual budget surplus and a rainy day fund of $550 million—and that in a state of only 1.7 million people. With the federal deficit in the trillions, those are trends that could grow on you.

258. Alligators: *Lungin' lizards*.
Once endangered but now more than 1 million strong, these powerful reptilian predators go back 150 million years. Alligators live in freshwater rivers, lakes, swamps, and marshes in southeastern states, mainly Florida and Louisiana. Males can grow to 15 feet and 1,000 pounds. Clumsy on

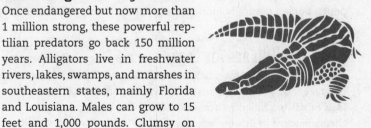

land, they reign in the water. Alligators like to dine on fish, turtles, snakes, and small mammals, but can stomach household pets and even humans too.

257. *It's a Wonderful Life*: Our Answer to A Christmas Carol.

It's officially the Christmas season when we gather around the television to watch George Bailey meet Clarence the Angel. We love it even though, upon reflection, we concede it is a really dark film about a man who contemplates suicide and believes the world would be better if he had never been born. Somehow, however, filmmaker Frank Capra's magical ending deludes us year after year into thinking of this 1946 Jimmy Stewart vehicle as an uplifting Christmas classic.

256. Beavers: *Testing our ability to coexist with nature.* Castor canadenis was on the brink of extinction in the 1940s but made a comeback. After humans, beavers change the landscape the most of all mammals, creating a man versus beaver conflict. Dam-building hurts trees and causes flooding disruptive to modern development, yet is beneficial to the preservation of wetlands. Can we just get along? Yes. Conservationists have invented high-tech "beaver bafflers" that mitigate the effect of Bucky being Bucky. If "preserving family values" is your no. 1 political issue, you will salute these ecologists' efforts. Beavers are rodents, but are naturally monogamous, unlike many politicians.

255. Seat Belts and Air Bags: *Grim Reaper's straitjacket.* The

slaughter of 42,000 Americans on our roads every year is a national disgrace. The good news: restraint systems save 4,000 lives a year. Half of drivers killed in accidents didn't buckle up, so please: join the 82 percent of drivers who do.

254. Whistleblowers: *J'accuse.* It's risky to allege misconduct by government or business when your livelihood is endangered. But heroes like Cynthia Cooper, head internal auditor at telecom giant WorldCom Inc., take that leap. Concerned over odd entries, Cooper and her team studied the books at night to elude detection. Although superiors repeatedly stonewalled her, she persisted. WorldCom was finally exposed in 2002 for $11 billion in fraud. In 1989, Congress enacted legislation to protect federal whistleblowers' jobs. Other ground-breaking laws followed including the No-Fear Act and the Sarbanes-Oxley Corporate Whistleblower Protection Act.

253. Ketchup: *The vegetable condiment.* President Reagan got tremendous flak when his administration categorized ketchup as a vegetable for the purposes of the federal school lunch program. Ketchup users, however, didn't mind. It was comforting to be told we were ingesting our daily serving of vegetables by slathering tomato ketchup over everything. If you don't like to eat it, you can still use ketchup as a hair conditioner or silver polish. For the record: we prefer Heinz and *never* store the bottle in the refrigerator.

252. Extension Services: *Agribizbuddies.* From their start in 1862 when the Morrill Act created land grant colleges, agencies that help farmers improve production and manage finances have been woven into the fabric of rural America. Wherever there's a puzzled farmer, there are experts like William Reid at Kansas State University to teach nut growers how to keep the pests off their pecans, or Vinicius Moreira at Louisiana State University, ready with tips on choosing hybrid corn silage for dairy cow feed.

251. Saddleback Church: *Wellspring of social justice.* Saddleback is the third biggest of the 1,360-plus megachurches in the United States. The pastor is Rick Warren, author of the best seller *The Purpose-Driven Life* and the clergyman chosen by President Obama to give the invocation at his 2009 inauguration, to the dismay of liberals for his stand against abortion and his opposition—later watered down—to homosexual marriage. But Warren has also been a tireless campaigner for increasing foreign aid, alleviating world poverty, and fighting the AIDS pandemic and global warming. Many megachurches, which run the gamut of 65 denominations (or else claim no affiliation), are self-contained villages, with day-care centers, schools, sports facilities, counseling services, social clubs, cafés, and restaurants. Most tap their deep pockets to try to create a better world, to the applause of their supporters and howls of their critics.

250. Walden Pond: *Eastern Eden.* This would be just another scenic tourist spot in Concord, Massachusetts, but for naturalist Henry David Thoreau's *On Walden Pond.* From 1845 to 1847, Thoreau lived in a hut on the pond's wooded shore. The beautiful locale inspired him to write his famous book, which helped raise Americans' awareness and respect for the natural environment. Many consider Walden Pond, now a National Historic Landmark, as the birthplace of the conservation movement.

249. Country Music: *Angst with a twang.* Songs of love lost. Of hardscrabble lives. Of pickup trucks, guns, and cheap booze. Okay, there's more to country music than those themes, but they always strike a chord in fans of this folksy genre, the most popular music form after rock/pop. Country evolved from the "fiddle tunes" of the southern Appalachians, but its foundation was laid in the late 1920s by Jimmie Rodgers, dubbed the father of country music, and the Carter family's harmonious mountain-gospel and folk sounds. Snubbed by the effete music world (despite the tremendous popularity of singing cowboy Gene Autry), country came into its own through the first country music awards show in 1966. Winners included Merle Haggard and Buck

Owens. Personally, we can't get enough of Willie Nelson, Johnny Cash, Lee Ann Womack, and Brad Paisley.

248. Weight Watchers: *Where they like to see less of you.*

The Weight Watchers organization, the original diet group established in 1963, estimates that a million people attend their meetings weekly. If each of them loses the recommended two pounds a week, we're talking a whale of a reduction in blubber.

247. Confronting Race Issues: *Seeking harmony.*

We don't push our problems under the proverbial carpet. Racial tensions in the United States have simmered off and on and boiled over more than a few times in the last 150 years. But in the 50 years since the civil rights movement began in earnest, Americans of goodwill have learned to live together and regularly work hard to overcome differences. It isn't perfect. We know that. But, as a society, we really want to change bad behaviors. The election of President Barack Obama in 2008 was a giant leap for mankind in sorting out racial differences. There are still setbacks. The erroneous arrest of Harvard Professor Henry Louis Gates in July 2009, for example, sparked a media blitz that ended only after Gates, Obama, and the arresting officer sat down for a beer. However, we celebrate the American willingness to admit we have a problem and our often awkward but well-intentioned attempts to make things better.

246. American Sign Language: *Giving the deaf a hand.*

No one can say for sure how many signs there are in ASL. An online dictionary counted up to 1,200 in 2009 and was adding more each month. There is no single international system. Like all sign languages, ASL is more than just a matter of hands and fingers. It uses body movements and facial gestures. For example, "If you are in serious pain, you need to show it on your face," the online ASL University says.

OY VEY!

245. Yiddish Expressions: *Jewish yeast for our melting pot's linguistic feast.* Oy, gevalt! To the nebbishes, nudnicks, schmucks, schlemiels, klutzes, and putzes out there who have the chutzpah to dismiss as bupkes or even dreck our claim that these Germanic gems enrich our melting-pot people's everyday conversation, we say stop kvetching and get your big fat schnozzle out of our business, you yentas. We're just a bunch of mensches doing our schtick, and if you don't like it, sue us, already. (Lawsuit, shmawsuit.)

244. Southernisms: *Our country's meatiest metaphors.* True, you have to think hard to make sense out of some of them, like, "Ah'm so hungry, Ah could eat the bark off a bear," and, "Ah don't chew my cabbage twice." Some can be crude, a tame example of which is: " 'Scuse my burpin' but my butt ain't workin'." The best are both vivid and knee-slapping, like, "Ah'm as frustrated as a one-legged man at a butt-kickin' contest," "He squeezes a quarter so tight the eagle screams," and, "It's so dry the trees are bribing the dogs." Yes, people really say these things.

243. Carnegie Hall: *World's favored stage.* You know the old saw: "How do you get to Carnegie Hall? Practice, practice, practice." Cute but true. Musicians everywhere consider Manhattan's Carnegie Hall their mecca. Since opening in 1891 with Tchaikovsky conducting the inaugural concert, philanthropist Andrew Carnegie's shrine to music has showcased great performers from Paderewski and Rachmaninov to Marian Anderson and the Beatles. Even the Steve Miller Band has graced this stage.

242. Children's Hospital of Philadelphia: *Where babies go to get better.* The Children's Hospital of Philadelphia (CHOP, as it is

affectionately called) was the first pediatric hospital in the United States when it was founded in 1855. Recognizing that children are not just miniature adults when it comes to medical treatment, the hospital established the first pediatric medical training program for doctors. Today, CHOP is ranked as the no. 1 hospital for children nationally. Among its many accomplishments is the creation of neonatal intensive care units.

241. Camping: *Listening to the silence.* Whether it's pitching a pup tent in the backyard, settling among the pines in one of the 5,000 state or federal campgrounds, or heading into the desert with a blanket, water, grub, and flint and steel to make a mesquite fire, camping is the antidote to the seductive comfort of modernity and helps us connect with the spirit of the pioneers.

240. Texas: *Outsized state for outsized egos.* Don't get us wrong; we love Texans' swagger. Houston oilman hands a $100 bill to a taxi driver. "Don't you have anything smaller?" the cabbie asks. Surprised, the Texan replies, "You mean they make 'em smaller?" We love the vastness of the Lone Star State, its cattle ranches, its  forests and plains, its oil. We love the 25 million people scattered from the Deep South to the Southwest, their drawls, their guns, their 10-gallon hats. We love Texas cowboys and the Texas Rangers and people like T. Boone Pickens and the late oilfield-fire fighter "Red" Adair. We love the history of Texas as a territory variously of Spain, France, and Mexico, as an independent republic governed by Sam Houston, as home of the Alamo and birthplace of the rodeo. Its capitol dome is the tallest in the U.S. Everything is bigger in Texas.

239. Niagara Falls: *Water, water everywhere.* Can't visit Victoria Falls in Africa? Niagara is the next best thing. Every minute, this natural phenomenon sends up to 6 million cubic feet of thunderous water over its crest to the swirling Niagara River below. The falls are 188 feet high, the rim of Horseshoe Falls on the Canadian side 2,200 feet across. Budget-minded honeymooners flock here. Publicity hungry daredevils still take the murderous plunge. First to try was schoolteacher Annie Taylor, who, in 1901, loaded herself and her cat in a barrel and let 'er rip. Miraculously, the 63-year-old and her tabby bobbed up alive.

238. Gettysburg National Military Park: *The high-water mark of Civil War sites.* A tour of this hallowed ground gives an emotional lesson on the three-day battle that turned the tide of the War Between the States. Seeing the long sloping hill to Cemetery Ridge where union cannons stood gives visitors instant understanding of why the climactic charge led by Confederate General George Pickett was doomed. Hiring private guides to the battlefield for $55 per vehicle is money well spent. These experts know everything about the strategy, the decisive moments, and how the 46,000 casualties gave "the last full measure of devotion," as Lincoln put it in America's greatest address. The house where he edited the 256-word speech before delivery was restored and opened for visitors in 2009.

237. Fall Foliage: *New England's color spectacular.* It's as though God Himself took out his paintbrush and gave the dying leaves of maples, oaks, and aspens a taste of immortality with His palette of burnt oranges, violent reds, and heart-breaking yellows and golds. Many Americans who live beyond the magic that has inspired untold songs and poems make annual pilgrimages to behold it.

236. Philadelphia: *History as we live and breathe.* Dubbed the City of Brotherly Love for its founder, Quaker William Penn, Philadelphia was the seat of federal government in the 1790s. In Old City, the treasured sites and symbols linked to the birth of our nation include Independence Hall, scene of the signing of the Declaration of Independence in 1776 and the framing of the Constitution in 1787. Across a cobblestone street is the glass pavilion sheltering the cracked Liberty Bell. Statesman Benjamin Franklin's grave attracts thousands at Fifth and Arch streets. Nearby is Flag House, where Betsy Ross created the Stars and Stripes. Modern Philly pleases too, with the grand 30th Street Station, the main railroad station, Reading Terminal Market—the oldest public farmers' market in America—and Rittenhouse Square, with its trendy shops, galleries, and restaurants.

235. Blueberries: *Antioxidant power-houses.* We loved these sweet to tangy berries in pies, pancakes, and muffins even before they were crowned heart-disease and cancer fighters. These native American berries, whose colors can range from blue to purple-black, are at their best from May through October. Total annual harvest: 536 million pounds.

234. Mickey Mouse: *Empire builder.* With his squeaky voice and enormous ears, Mickey wowed 'em in the first sound cartoon, *Steamboat Willie*, in 1928. His success spawned other cartoon characters: Minnie Mouse, Clarabelle Cow, Goofy, Pluto, and Donald

Duck, among others. Mickey starred in the animated feature *Fantasia* and inspired television's *Mickey Mouse Club*. Today he's an international symbol, personifying everything Disney.

233. Tap dancing: *Oh, those crazy feet!* An American innovation of the 19th century, tap dancing is still popular today. People love doing the shuffle, shim sham shimmy, and heel clicks. Named for the tapping sounds made when metal plates on dancers' shoes strike the floor, tap's roots go back to English, Irish, African, and West Indies dances. Attention all hoofers: National Tap Dance Day is May 25, the birthday of famous tapper Bill "Bojangles" Robinson.

232. The Merit System: *It is what you know.* Most Americans believe that their government runs best when public servants are bright and hardworking. The old-boy network. Nepotism. Affirmative action. Veterans' preferences. Politics. These hooks get people in the door for public jobs, and sometimes get them promoted. But brains and skill are required for most of those who rise to the top. The merit system took root in the 1880s with the civil service movement, and many states now have laws that require public bodies to hire the top finishers on civil service exams. For many jobs on the local level the mayor's pals may have the edge. Being qualified might save them when the mayor is voted out.

231. Arlington National Cemetery: *Consecrated ground.* This 200-acre Virginia tract across the Potomac River from Washington, D.C., is the final resting place for more than 300,000 of our military. Veterans of all wars from the Revolution through Iraq and Afghanistan are buried here. Many of the 4 million annual visitors pay their respects at the Tomb of the Unknowns, where three unidentified

American soldiers killed in two world wars and the Korean conflict are interred. (Remains of the Vietnam War unknown were exhumed in 1998, then identified through DNA.) Down a slope from the Custis-Lee Mansion, an eternal flame marks the grave site of President John F. Kennedy, assassinated in November 1963. Some eight months earlier, Kennedy had toured the mansion, observing that its view of Washington was so magnificent, he could stay there forever.

230. Trying to Win the Mega Millions:

It could happen! The proceeds of lotteries helped fund the Jamestown settlement and equip our Revolutionary Army. Lottery is big business—it supports about 250,000 jobs and brings profits to the 240,000 retail outlets that sell tickets. We include playing the Powerball and Mega Millions lotteries (sold nationally in more than 30 states) because Americans actually think they might win!! Chalk it up to our sense of optimism, since you only have about one chance in 195 million of grabbing the top prize. Hats off to Helen Lerner of Rutherford, New Jersey, who held the only winning ticket for a Mega $258 million jackpot.

229. The Lincoln Memorial: *Temple of freedom.* The imposing 19-foot high statue of Lincoln, the lack of adornment, the grand scale, and the view across the Washington Mall to the Capitol make visitors feel like they are in a great cathedral. Since it was dedicated in 1922, the memorial has been a place for contemplation and the occasional outpouring of emotion, as in 1963 when Martin Luther King, Jr., declared there, "I have a dream." President Richard Nixon made a visit before dawn on May 9, 1970, to speak with anti-Vietnam War protesters. "That's all right," he told them. "Just keep it peaceful. Have a good time in Washington and don't go away bitter." Not too eloquent, but next to Lincoln, who is?

228. The Union: *Now and Forever.* The South disputed the supremacy of the people, and its defeat by the North in the Civil War forever settled the question of who is boss (despite occasional grumblings in this state or that). "The Union" might not roll off most lips as one of the greatest things about our country, but it should. It was paid for by the blood of 620,000 dead—out of a population of 31 million.

227. *The Sun Also Rises*: *Poetry in prose.* Ernest Hemingway's first novel made him a literary sensation. Although published in 1926, the book's terse sentences—the hallmark Hemingway style that influences writers to this day—makes it oh so evergreen. This lost generation novel has a cast of boozing expatriates who, except for the impotent narrator, bed or battle one another across France and Spain.

226. Hollywood: *International sign of glamour and excess.* If you want to be an international film star, move to California's Tinseltown, where the 50-foot-tall "Hollywood" sign perched on a nearby mountain welcomes you to one of the toughest businesses in the world. It's the city where, as Burt Bacharach wrote, "dreams turn into dust and blow away." Hollywood is a magnet for ingenues and tourists alike. You can watch movies filmed at Universal Studios, follow the Hollywood Walk of Fame, and also take a peak at how the few who make it big actually live. Take a bus tour of celebrity homes in Beverly Hills or loiter in the shopping district on Rodeo Drive. If you're lucky, the stars may come out.

225. Spy Satellites: *Eyes in space.* As a rule, we're not fans of expensive weaponry, but satellites that can spot a budding missile system in a rogue state or listen to a potential enemy's radio signals make the world safer and give us confidence to reduce nuclear arsenals.

224. Magazine Mania: *Window on American culture.* Sure, a lot of mainstream magazines have been hurt or knocked out of the market by the Internet and a flailing economy. But others find a winning formula in responding to the latest trends. The number of home magazines almost tripled to 241 in 2008, from 83 five years earlier, and publications about winter sports doubled to 106, though regional magazines held top spot with 1,126 titles, including *Garden & Gun*. Home and garden magazines have also seen a resurgence. Trade magazines prosper. The glossies' poor cóusins, newsletters, though they focus mostly on practical matters like the law and computers, also go as far afield as instructing consumers on how to find the best pet doors.

223. The Farm Labor Movement: *Genesis of Latino political power.* When Delano, California, grape growers cut pay rates, laborers struck on September 8, 1964. They soon were joined by the National Farm Workers Association—later to become the United Farm Workers Union—with Mexican-American organizer Cesar Chavez in charge. Through nonviolent tactics, including a nationwide boycott of nonunion grapes supported by 13 million consumers, most table-grape growers were signed on by 1970. Over the years, there have been some setbacks in fighting for their cause, but the UFW has achieved historic gains for struggling workers, including a collective bargaining agreement and union contracts requiring rest periods, field toilets, clean drinking water, and regular testing for pesticide exposure. Lately the union has been at the forefront of the illegal immigrant debate, urging Congress to pass legislation allowing undocumented laborers to gain legal status provided they continue working in agriculture.

222. Our *Carpe Diem* Attitude: *Seizing opportunities.* Sometimes you just need a good idea in the right place and at the right time. Think of childhood entrepreneurs who sold lemonade on hot days then grew up to market umbrellas on city streets in the rain. We celebrate the guy who made a buck hawking HELP! OUR PRESIDENT IS A MORON T-shirts during the last decade, and the folks who capitalized on President Obama's election by selling T-shirts emblazoned with his face.

221. Tailgate Parties: *Flame before the game.* Nothing is more American than burgers, beer, and loud music amid a sea of automobiles, preferably big ones. NASCAR blogger Steve McCormick gives a high rating to the tailgate parties at the 400-mile race at Dover International Speedway in Delaware. "Where else can you see a guy dancing naked to banjo music in the middle of the night while pouring beer on himself?" On the other end of the spectrum: to qualify for scarce space at the party before the 2006 Yale-Harvard football game, Harvard created selection criteria almost as rigorous as the school's undergraduate admissions process. And booze was banned.

220. Third Parties: *Sideshows that keep our two-party system honest.* Their aims have ranged from lofty to loathsome, but the Big Two (in our day, the Democrats and Republicans) have always been quick to coopt their causes. Three contenders got on the ballots of enough states to qualify as third parties in the 2008 presidential election: the Libertarian Party, the Constitution Party, and the Green Party. As for the countless flameouts of yesteryear, we highlight their names alone for the antiimmigrant No-Nothing movement of the 1850s whose adherents were instructed to say "I know nothing" about its activities, and the Bull Moose Party, a 1912 Republican splinter group that gained its moniker from founder Theodore Roosevelt's boast he was "strong as a bull moose."

219. New Orleans: *All that jazz.* A boisterous mix of European traditions and Caribbean influences, "The Big Easy" seems more for-

eign than American. Founded by the French in 1718, New Orleans anchors the world's busiest port complex: Louisiana's lower Mississippi River. With French and Spanish street names, sherbet-hued buildings, spicy Creole cuisine, and omnipresent jazz, this city's zest for life continues despite the devastation of Hurricane Katrina and dogged attempts to rebuild. *Allons-y, laissez les bon temps rouler!*

218. Volleyball: *That other sport invented in Massachusetts.* We all know about basketball's auspicious beginnings in Springfield, Massachusetts, but a little known fact is that four years later, in 1895, and 18 miles up the road, volleyball was

born. It must have been really boring in New England in the winter to spawn two such active indoor sports. Thanks to those cold and wet months, we now celebrate volleyball as an important collegiate and Olympic sport.

217. Pickup Trucks: *Fully loaded, or not.* Whether it's a beat-up 1988 Ford F350 bearing rust and the smell of manure or a compact new Suzuki Equator parked in the garage of a city apartment, a pickup truck tells a lot about its owner. Americans nursed their aging pickups through a bad economy in 2009. Sales were down 30 percent from the previous year. The most popular model, with 400,000 buyers, was the Ford F-series. We will avoid the argument over which company designed and sold the first pickup. But we can say for sure that the early versions that traveled the muddy roads of post-World War I America were flimsy, inexpensive vehicles for farmers and tradesmen, not the icons of rural chic that pickups have become. Like the $115,000 International CTX, the largest production pickup on today's market.

216. Thinking We're Top Dog: *Love us or hate us, the world has to envy us.* We have won more Nobel prizes than any other country (eight in 2009 alone, including the controversial Peace prize

awarded to President Obama). We have the largest gross national product, the mightiest military, the best research labs, the most advanced technology, the most sought-after universities, the most innovative music, art, and film; the most entrepreneurial culture, the greatest ethnic diversity, and the most productive farms of any nation on earth. We whupped the Brits, Nazis, Japanese, and Soviets in wars hot and cold. We created and fixed the Y2K threat. We were the first and only people to walk on the moon, and that was four decades ago. We bestride the continent, from sea to shining sea. Hell, we bestride the globe. President Reagan spoke of us as a shining city on a hill, blessed by God, whose "glow has held steady no matter what storm." *We're no. 1.*

215. Our Humility: *The perfect antidote to our boastfulness.*
The British, Germans, French, Italians, and Dutch together have won nearly as many Nobel prizes as we have, with a smaller combined population. With Europe's currency challenging the greenback, China's and India's economies gaining on ours fast, and Russia (twice our geographic size) reasserting itself—to say nothing of pipsqueak dictators everywhere poking us in the eye—we are undergoing one of our periodic spasms of angst over our inevitable decline. Besides, we *feel* inferior to much of the world. The British are more urbane, the French better lovers, the Latin Americans more fun-loving, the Italians more passionate, the Israelis more inventive, and the Egyptians and the Greeks endowed with a profounder sense of history. *Gads: we won't be no. 1 much longer.*

214. The American Musical: *Of thee we sing.* Rodgers and
Hammerstein's 1943 musical, *Oklahoma!,* was the first stage vehicle to fully integrate book and music. This melding of dialogue and songs, with the music also serving to advance the plot, is a distinctly U.S. art form. We know it through classics like *South Pacific, The King and I, West Side Story,* as well as *Hello, Dolly! Fiddler on the Roof,* and *Cabaret.* Then Stephen Sondheim teamed with director Hal Prince to redefine this form, resulting in *Company, Follies* and A

Little Night Music—cutting-edge shows that defy traditional notions of plot. That's great! And now, to draw larger audiences in this era of economic need, musicals are incorporating rock, country, and pop-soul sounds. *Jersey Boys* and *9 to 5* come to mind. Fine! Just as long as they remember the old litmus test: a successful score has at least three "hummable" songs.

213. Surfing: *The ocean's waves, not the Internet.*

We trace the history of surfing to the Hawaiian art form, *he'e nalu,* making it an American original. Surfing surged in popularity after the 1959 Sandra Dee movie about teenage surfer girl *Gidget.* From May to September the world's best surfers congregate by the 20-foot-plus waves on Waikiki Beach, Hawaii. Hang 10!

212. River Cleanups: *Work against the murk.*

From the U.S. Environmental Protection Agency with its billions of dollars and legions of scientists and lawyers, to the volunteers in rowboats who pick up floating Budweiser cans, America is cleaning up its rivers one eddy at a time. Big rivers like the Mississippi still serve as sewers. But there are success stories. Like the Detroit River, where the whitefish are coming back, and like the newly cleansed streams near Johnston, Pennsylvania, thanks to the Stonycreek-Conemaugh River Improvement Project and $5.6 million in government funds.

211. U.S. Postal Service: *In it for the long haul.*

Its 650,000 employees process 700 million pieces of mail *daily,* almost as much as the rest of the world combined, and do it efficiently and, usually, cheerfully. You can send a letter from New York to Los Angeles first-class for a fourth the price of a cup of coffee. Yes, the junk solicitations can drive you nuts, and the other envelopes contain mostly bills, but you never know: sometimes the check really is in the mail.

210. Watermelon: *Portable dessert.* It originated in Africa, was first cultivated in the Nile Valley in the second millennium B.C., and is grown mostly in China. So why is watermelon a must at every American barbecue? We eat it, for one thing, slurping down 25 pounds per capita per year. The Chinese turn it into oil and munch the seeds. Also, it appeals to our love of bigness (one weighed 262 pounds), sense of fun (the seed-spitting record is 75 feet and two inches), and obsession with health (the fruit has antioxidants and vitamins A, B, and C galore).

209. Los Angeles: *City of Angels and devilish traffic snarls.* L.A. has everything from the prehistoric La Brea Tar Pits and the parklike campus of UCLA to starstruck Sunset Boulevard, Beverly Hills's luxurious Rodeo Drive, and Little Tokyo. The 44-acre El Pueblo de Los Angeles urban park includes early structures, a busy Mexican market, and Avila Abode, said to be the oldest standing structure in this city founded by the Spanish in 1718. As home to Hollywood, L.A. also is known as the "Entertainment Capital of the World," for the many films, television shows, and recorded music produced there. Greater Los Angeles has nearly 12.9 million inhabitants, second only to New York City. Alas, L.A. tops New York and other U. S. metro areas in traffic problems, with delays often extending from sunrise to sunset.

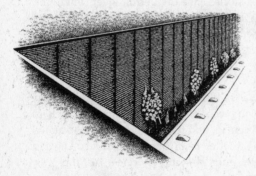

208. The Vietnam Veterans Memorial: *Hallowed walls.* This memorial on the National Mall in Washington, D.C., honors the 2.7 million who served in the Vietnam War zone. Its heart-wrenching black granite walls are inscribed with the names of more than 57,000 dead and 1,200 still missing. Designed by then-Yale undergrad Maya Ying Lin, the walls were dedicated in 1982.

207. J. Paul Getty Museum: *West Coast answer to New York's Metropolitan Museum.* When you are the world's richest dude, you can get a museum named after yourself. In life, oil billionaire Getty was so cheap he had pay telephones installed in his house; in death, his billion-dollar legacy provided the world free access to an amazing art collection housed in two sites, in L.A. and Malibu. Its extensive holdings, along with its location on the breathtaking California coast, make it one of the world's greatest museums.

206. Florida: *A taste of the Tropics.* Legend says Spanish explorer Ponce de Leon went there in 1513 to find the Fountain of Youth. Millions of cold, retired Americans have followed his example, making Florida a mecca for the wrinkled set. "My parents didn't want to move to Florida, but they turned 60 and that's the law," comedian Jerry Seinfeld joked. More than a land of condos and cabanas, Florida has become a haven for Hispanics, mostly from Cuba, who represent 11 percent of the electorate statewide and 52 percent in Miami-Dade County. Since the election of 2000, when the state's disputed vote was the margin that made George W. Bush President, Florida has been a crucial political swing state. When it comes to citrus, there's no contest. Florida oranges win our vote over California's.

205. Mount Rushmore: *The Great Pyramids of America.* Gutzon Borglum completed the colossal granite sculptures of Lincoln, Washington, Jefferson, and Theodore Roosevelt in the Black Hills of South Dakota in 1941 after just 14

years of toil. But isn't that what America is all about: speed? Tourism boosters declare Mount Rushmore to be both a shrine of democracy and "the greatest **FREE** attraction in the U.S." For all the marketing blather, the rock carvings nearly match the magnificence of the only remaining Great Wonder of the Ancient World.

204. The Red Cross: *Looking out for you.* Part of the International Red Cross and Red Crescent Movement, the American Red Cross was founded in 1881 by Clara Barton. As the nation's premier emergency response organization, it boasts more than 700,000 volunteers, 34,000 employees, and nearly 720 locally supported chapters. In addition to disaster relief, the Red Cross also supplies blood and blood products.

203. Mom's Meat Loaf: *Real home cooking.* Meat loaf, a staple of American cooking, only tastes good if it's *your* mom who made it. Very few moms can give you a recipe. It's usually ground meat, with a dash of this and a touch of that. Some Moms shred carrot and zucchini in as a sneaky way to feed their kids vegetables; some add Tabasco for kids who need a wake-up call, while others have been known to sprinkle powdered fiber into the mix to regulate their children's bowels.

202. Triple-A Baseball: *Affordable professional sports.* In 2009 a family of four paid an average of just $62 for tickets, hot dogs, soda, beer, and parking at Triple-A games, whose players are the crème de la crème of the Minor Leagues. That was less than one-third the comparable cost of $197 for a night out at a Major League stadium. No wonder attendance in Triple-A ballparks, in cities from Sacramento to Durham, North Carolina, rose to 14.4 million in 2008, from 9.5 million in 1990.

201. Oscar Night: *Hollywood's annual self-love fest.* The statuettes are gold plate, much better than the plaster ones bestowed during World War II. No one can say for sure who first called the Academy of Motion Picture Arts & Sciences awards Oscars, but the name stuck in 1934. Walt Disney won a record 26 times. The awards ceremony is often TV's highest-rated live entertainment show of the year—55 million people watched *Titanic* win in 1998—though

many viewers now prefer the preshow red carpet parade where stars display their fashions and facelifts. Six of the movies on the American Film Institute's all-time top 10 list won Oscars for best picture, which suggests the Academy's voters have a pretty good record of getting it right. Too bad the shows are so long that legions of viewers on the East Coast are asleep before the final Oscar is awarded.

200. Gone Fishin': *The big escape.* Americans work 1,800 hours a year, far more than inhabitants of most industrialized countries. So when we tack a GONE FISHIN' note to our desks or inform callers by voice mail that "I'll be away from my desk until Monday, March 15," we are advertising a talent no other nation can match: serious goofing off.

199. Boston: *Our nation's birthplace.* Puritans seeking religious freedom established a settlement above the Charles River in Massachusetts in 1630. By the late 1700s, Boston was a bustling port. But less than a century later protests over British rule, notably the Boston Tea Party, helped spark the American Revolution. Now New England's largest city, Boston teems with students. The area hosts some 60 colleges and universities, including Harvard, MIT, Boston U, and Northeastern, giving Boston the highest concentration of students of any major U.S. city. Downtown is anchored by Boston Common, the nation's oldest public park. Nearby are historic row houses on Beacon Hill, and Faneuil Hall, a favorite marketplace and meeting hall since 1742. Dubbed "Bean Town" for its famous molasses baked beans, Boston's ethnic heart beats in the picturesque North End. Currently called Little Italy, it's been home to successive waves of immigrants, especially Irish.

198. *The Night Before Christmas:* The poem that solidified Santa Claus. Clement Moore (1779-1863) was a professor of classics at New York General Theological Seminary. As such, he probably would have preferred to be immortalized for his two-volume *Compendious Lexicon of the Hebrew Language.* Not to be. As legend has it, he wrote the most recited poem of all on a sleigh ride home to Greenwich Village in 1822. He envisioned Santa as "chubby and plump, a right jolly, old elf," and the image stuck.

197. Local Hardware Stores: *Thriving in a big box culture.* You walk in and ask for a can of paint, a gas hose adapter, or a toilet tank repair kit, and a weird thing happens: a human being gives you what you want. And advises you how to use it. Better to pay a premium for such service than to wander in the aisles of Lowe's or Home Depot, crying like children lost in a midnight forest. "Hello. I need a flush valve seat. Can anyone hear me? Is anyone there? Heeeelp."

196. Polio Vaccines: *Making the world safer for kids.* Anybody over 60 remembers the dreaded summer polio epidemics that closed swimming pools and condemned thousands of victims to wheelchairs, iron lungs, or even early deaths. Then vaccines developed by Jonas Salk in 1952 and Albert Sabin in 1962 led to a spectacular decline in cases and to hopes for a worldwide eradication of the disease. Famous polio survivors include actor Alan Alda and President Franklin D. Roosevelt.

195. Labor Unions: *Watchdogs for the underdogs.* We have union activism to thank for establishment of the Department of Labor in 1913 to protect wage earners' rights and welfare. American workers also successfully won the right to organize, engage in collective bargaining, and strike. Along the way, however, the labor movement has suffered wrenching strikes, internal corruption, and weakened clout. This once powerful force, which represented 35 percent of the nation's workers by the 1950s and helped create the world's largest and richest middle class, now represents just 7.4 percent of private-sector workers. Nonetheless, the movement perseveres, fighting for higher pay, better health care, pensions, and restored collective bargaining rights. Some think that with our struggling economy, workers losing even more ground, and corporate executives still pocketing huge bonuses, we might be heading for a much-needed rebirth of the labor movement.

194. Chicago: *City of big shoulders.* "Come and show me another city with lifted head singing so proud to be alive and coarse and strong and cunning," poet Carl Sandburg wrote in 1916. Chicago has calmed down since then. It is considered one of America's best-run cities, despite—maybe because of—its machine politics. No longer hog butcher to the world or the crime capital, Chicago is the business and cultural center of the Midwest, has stimulating concerts, museums, restaurants, sports teams, universities, and a public transportation system that rivals New York's. The skyline does too. The 1,450-foot Willis (formerly Sears) Tower is North America's tallest building. Writing about Chicago in 1951, novelist Nelson Algren said, "Like loving a woman with a broken nose, you may find lovelier lovelies, but never a lovely so real." The city remains real, but the face is prettier.

193. The Smithsonian: *America's attic.* It's a treasure trove of really cool things that we as a people couldn't bear to throw away. Visitors to one of the 19 venues that encompass the Smithsonian Institution in New York City and Washington, D.C., can see rocker Prince's electric guitar, jazz man Dizzy Gillespie's trumpet, daredevil Evel Knievel's Harley-Davidson, and other artifacts of American culture at the National Museum of American History. The ever-popular Air and Space Museum chronicles our fascination with the galaxy, while the National Zoo's Amazonia Rain Forest brings the jungle to us.

192. Popcorn: *Kernels we salute.* Popcorn was in the native diet when Spanish explorers arrived in the 16th century, and its popularity as a snack food has exploded like a popped kernel. Some of us go to the movies just to get a jumbo portion drenched in hot butter. FYI to dieters: Consumer Search gives its best rating to Orville Redenbacher Smart Pop 94 percent fat-free butter. The Popcorn Board estimates that the average American consumes 54 quarts of popcorn a year. Good news for the dental floss industry.

191. Interfaith Cooperation: *The God Squad.* That is the name of a syndicated column in which a Catholic priest and a rabbi respond to readers' questions about faith and morals. We use it to signify all the organizations and movements, from Evangelicals and Catholics Together, and the International Fellowship of Christians and Jews, to the Center for Jewish-Muslim Relations, that thrive in our pluralistic society. One of our favorite billionaires, John M. Templeton, the mutual-fund maestro who died in 2008, showered tens of millions of dollars on Christians, Jews, Hindus, Buddhists, Muslims, and even agnostics who fostered understanding about ultimate "spiritual realities."

190. The John F. Kennedy Space Center: *Somewhere, out there.* Since 1962 this Florida spaceport of the National Aeronautics and Space Administration has been the launch site of every manned U.S. space mission and hundreds of exploratory spacecraft. NASA was created partly to meet the challenge of the Soviets' first artificial satellite in 1958. Ever since, our unending curiosity about the universe has compelled us to launch satellites and space shuttles, send the Mars *Pathfinder* to search for life on that planet, and establish a permanent human presence aboard the orbiting International Space Station, a 16-nation project. The trajectory has been both triumphant, with Neil Armstrong's walk on the moon, and so-

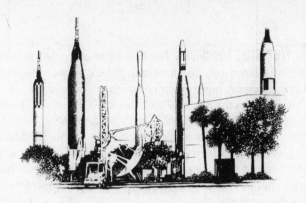

bering, with the deaths of 24 astronauts in space-related accidents or aircraft crashes.

189. Americans with Disabilities Act: *Legislation that opened doors and made them wider.*

The United States began addressing civil wrongs in earnest in the 1960s, but it wasn't until 1990 that Congress addressed the needs of 43 million Americans with some form of disability. The law required handicap access for the physically disabled, equal access to education, recognition of mental problems as disabilities, reasonable accommodations at school and work, and equal rights in the job market. In short, the legislation reminded us that Americans with disabilities are people too—with the same aspirations, goals, and dreams that everyone else has.

188. The "Gourmeting" of America: *Very choice chow.*

After World War II, meat-and-potatoes America grew up. Not overnight, mind you. It took decades for us to rise above Campbell Soup casseroles, oatmeal-laced meat loaves, and breaded fish sticks as dinnertime staples. If now we view risotto, coq au vin, or rib-eye steak with chipotle glaze as practically our birthright, we can thank our growing affluence, our increased travel to culinary-rich foreign lands, and the astounding popularity of cookbooks and TV cooking shows. Toques off to the gourmet gurus who helped us develop these discerning palates. Let's hear it for chefs James Beard, Julia Child, Alice Waters, Todd English, and John Ash, among many.

187. Warming the Bench: *Life lessons at an early age.* Life for most of us is laced with disappointments, so it's good to learn early to deal with failure. Youth sports is our school for stoicism. We learn that our fate is in the hands of God: the coach who makes his own, untalented kid quarterback while we ride the bench. We find out that the line between winning and losing is as thin as volleyball netting. We accept that referees will stop calling fouls in the final quarter because their dinner is getting cold. If we are left-handed we learn that we will never play catcher, second, short, or third. Life is hard. We move on.

186. Volunteer Coaches: *Helping kids have a ball.* Until the 1950s most children who wanted to play a sport met their pals and chose up sides. In these days of organized youth sports, where would youngsters be without the hundreds of thousands of dads, moms, siblings, and grandparents who volunteer to teach skills and sportsmanship, manage the games, and serve as referees and umpires? They do it for the thrill of turning a butterfingered fielder into a kid who just might have a chance of making the putout that wins the game. The hard part is dealing with parents who wonder why their future Michael Jordan or Mia Hamm is warming the bench.

185. The Peace Corps: *Cultural bridge to the world.* Since the agency was created in 1961, more than 200,000 volunteers have ventured into the developing world for two or three years to teach English, design buildings, dig wells, and the like. Immersing themselves in the local culture, they have become an army of grassroots diplomats. The most famous was President Jimmy Carter's mother Miss Lillian, who joined in 1966, at age 68, and spent two years in India as a nurse. Other returned volunteers include Netflix founder Reed Hastings, travel writer Paul Theroux, and former Ohio Governor Robert Taft. Today, nearly 8,000 adventurers carry on the tradition in more than 70 countries, branching into such areas as AIDS awareness programs and information-technology training.

184. The U.S. Marines: *Lethal eagles*.

Call them leathernecks, jarheads, or Devil Dogs, they project American power to foreign shores. Their hymn, which opens with "From the Halls of Montezuma to the Shores of Tripoli," celebrates victories in the Mexican-American War and the campaign against the Barbary pirates. They raised the flag on Iwo Jima in World War II and conducted operations against the Taliban and Al Qaeda after 9/11. In between, American Presidents obeyed the dictum "Send in the Marines" more than 200 times. The Few. The Proud. The Marines. Thanks, folks.

183. Maine Lobsters: *Clawing to the top*.

Down Easters tout their coast as ideal for nurturing the finest lobsters. There's cold, clean water and a rocky bottom habitat where these spiny creatures feast on crabs, mussels, and other ocean delicacies. The show-stopper: new-shell lobsters. Annually, mature lobsters shed old shells for new, larger ones. The meat of these newly encrusted crustaceans is superflavorful, with shells that can be cracked by hand.

182. Moose: *King of the beasts of North America*.

They are big, ugly, and look like they need a shave. They can swim and run, but waddle when they walk. Despite that, with a crown of antlers that can span nearly seven feet, moose project a magisterial dignity that no other animal on our shores can match. For that we pay them homage. Moose can be spotted in the northern states, most especially Maine on the East Coast, Minnesota in the Midwest, and, in the West, in Alaska, which honors the mighty moose as its state mammal.

181. Golden Gate Bridge: *Engineering as art.* Consulting designer Irving Morrow rejected the classic steel gray color and picked International Orange, whose warm tones blend with the land on either side of America's most beautiful major span, which links San Francisco with suburban Marin County. Aesthetics be damned, the bridge authorities decided in 2008 to put steel netting under the roadway to stop suicides—1,300 and counting since the bridge was finished in 1937.

180. The Fifth Amendment: *The Constitution's high five.* Governments are by nature power-hungry, but the five lofty provisions of this amendment defend us from legal abuses by ours. They stipulate the rights to a grand jury and due process of law, and protections against double jeopardy, self-incrimination, and the taking of property without fair compensation. We may blanch when mobsters or financial hucksters "take the Fifth" on the witness stand, or when criminals literally get away with murder because they can't be put on trial a second time on the same charges, but that is the price we willingly pay to keep overzealous prosecutors in line.

179. Harvard University: *High-profile pipeline.* Established in 1636 in Cambridge, Massachusetts, our oldest institution of higher learning attracts the most accomplished students. Harvard's academic prowess is stunning, with 43 Nobel Laureates among current and former faculty members. Noted graduates include five of the nine sitting Supreme Court justices (also, luminaries Oliver Wendell Holmes, Louis Brandeis, and Felix Frankfurter), and eight Presidents, among them FDR, Kennedy, George W. Bush for his MBA, and Barack Obama in law. Considering this company, a degree from our own Oxbridge has cachet bar none.

178. Big Pharma: *Purveying medicine from Advair to Zantac.* The U.S. pharmaceutical industry earns more than $300 billion a year and spends one-fifth of it on research and development to churn out more remedies to treat ailing bodies from hair (Provcillus[1]) to toe (tolnaftate[2]). Drug companies get a bad rap every time they must fork over billions of dollars because a product is suspected of sickening or killing people, like Vioxx, the painkiller recalled when it was linked to heart disease. But the good stuff overshadows the bad. Tylenol and Advil, for example, are great for headaches caused by trying to read the small print on pill bottles.

177. U.S. History: *Short and sweet school subject.* Pity those poor Chinese kids whose national history books start around 1100 B.C., or the Italian kids who need to memorize all the Caesars in the fourth grade, or the Greek waifs who have to sort myth from reality. Better to be an American schoolchild, charged to learn a relatively short history that really doesn't heat up until the mid-18th century.

176. The Right to Bare Arms: *Women's fashion freedom.* In many parts of the world, especially in Islamic theocracies, modest dress for women is required by law or custom. In America a woman's right to show skin is constitutionally protected freedom of expression, limited only by workplace and school dress codes and the ever-shrinking view of what should be covered. The epidermal revolution started slowly in the 19th century, when naked wrists and ankles stopped shocking. Off and on since then, mostly off, necklines have plunged and hemlines have soared. Tummies and butts are the newest frontiers. Who knows how far it will go? Perhaps to ultimate freedom, which folk rocker Kris Kristofferson defined in a 1969 song as "another word for nothing left to lose."

---

1. Unico Enterprises hair growth treatment
2. Active ingredient in Dr.Scholl's athlete's foot cream

175. Silicon Valley: *Hotbed of high technology.* Located in the southern half of the San Francisco Bay area, Silicon Valley—named after the silicon-chip innovators that sprouted there in the 1970s—is to computers what Hollywood is to movies and Manhattan is to finance. Incubator of cutting-edge companies like Apple, Google, Intel, Yahoo, and Pixar, the region has redefined American folklore. Though other cities like Seattle and Boston have become high-tech magnets, Silicon Valley remains the big enchilada.

174. Free, Fair, and Frequent Elections: *Lifeblood of democracy.* All those hair-breadth-close margins—*George W. Bush* v. *Al Gore* in Florida in the 2002 presidential contest and *Al Franken* v. *Norm Coleman* in Minnesota in the 2008 Senate race come to mind—remind us that every vote counts and every vote is counted. Well, usually. Accusations of fraud have abounded in our history, but at least people are free to squawk, the press to poke around, and prosecutors to investigate. Contrast that with the sullen acquiescence or explosive violence that greets victories for dictators in rigged elections in places like Algeria, where President Abdelaziz Bouteflika won reelection in 2009 with 90 percent of the vote, or preinvasion Iraq, where Saddam Hussein nailed down 100 percent.

173. Centers for Disease Control and Prevention: *Wellness watchdog.* Since 1946 this government agency has worked to improve public health. Much credit goes to the Atlanta-based centers for the 25-year increase in Americans' life spans. Two key factors in this: widespread vaccination to eliminate polio and rubella while controlling other infectious diseases, and antismoking campaigns coupled with early detection and better treatment to produce a 51 percent drop in fatal heart attacks and strokes. The CDC often investigates outbreaks of foodborne illnesses. Recently it identified the bacterium salmonella in peanut butter as the culprit behind nearly 700 food-poisoning cases—at least two fatal—in 46 states. The contamination was traced to a Georgia plant.

172. Title IX: *Our attempt to level the playing field.* Although Title IX of the Education Amendments of 1972 prohibits gender discrimination in *any* educational activity, its biggest impact has been in sport locker rooms. In fact, it created an emphasis on girls in sports that simply did not exist before. While educational athletic programs still spend more on male sports, and media outlets still emphasize the men's teams, Title IX created vast opportunities for women. Eight times more girls are involved in high school athletics today than in 1972. Still, true parity is elusive. Women coaches earn less, and men who coach women's teams also earn less, just to name two injustices.

171. YouTube: *The global video village.* Founded in 2005 and owned by Google, YouTube is like television in 1949: big and getting bigger. Anyone with a video camera can be a producer, and anyone who wants 15 minutes of fame—usually it's more like 15 seconds—can be on the Internet. In early 2009, Chief Executive Officer Chad Hurley boasted that 15 hours of new videos were being uploaded to the site every minute, from Aab (a Danish football team) to ZzZ (a Dutch rock band). In 2009, singer Susan Boyle's "Cry Me a River" performance on *Britain's Got Talent* attracted 100 million visits in its first nine days on YouTube. Among our favorite posts: the video *Charlie Bit Me,* Bush and Obama bloopers, and William Hung's "She Bangs" audition on *American Idol.*

170. Fast-Food Deliveries to Your Door:

Fancy culinary footwork. Whether Chinese cuisine or southern-style ribs, the grub that is just a phone call away is a boon for just about anybody who doesn't live in the wilderness. That includes insomniacs, weary commuters, and the vast population of the cooking-challenged.

169. FBI: *Buster of bad guys.* For more than a century the Federal Bureau of Investigation has tackled everything from the mob, spies, and public corruption to civil rights crimes, bank robberies, and terrorism. Think Gotti, Enron, and September 11, for starters. Their mission never ends: in 2009, FBI negotiators worked to free American ship captain Richard Phillips from Somali pirates.

168. California Wine: *Nectar of a continent.* Spanish missionary Father Junipero Serra planted grapes in California in the 1770s. They flourished until a sap-eating insect ravaged the wine industry in the 1890s. Then came Prohibition in 1919. By the time legislated teetotalism ended in 1933, California's wine industry was in ruins. It took 40 years to recover, but recover it did. Americans first had to be convinced that wine was a sophisticated alternative to the liquor and beer we preferred. Today, the California wine industry—with a 61 percent share of the U.S. wine production—is a $19 billion business that annually sells 457 million gallons worldwide. Yes, you can even buy a nice California chardonnay in France. And a visit to the wine country in the Napa and Sonoma valleys north of San Francisco is the second most popular tourist destination in California, behind Disneyland.

167. Boy Scouts: *Teaching boys everything men need to know.* In 2009, Andrew Schigelone of Lincoln Park, Michigan, finished earning all 121 merit badges, from American Business to Woodworking. Most of America's 900,000 scouts are content to get outdoors, absorb the basics of survival and citizenship, and learn to tie a sheepshank knot. They are preparing to be prepared. The Scout law requires members to be trustworthy, loyal, helpful, friendly, courteous, kind, obedient, cheerful, thrifty, brave, clean, and reverent. Good qualities to maintain in adulthood. Scouting does not preclude boys from being boys. When scoutmasters weren't watching, the boys at Ten Mile River Scout Camp in New York in the 1950s and 1960s smoked an occasional cigarette, played poker, and exchanged misinformation about women. The tradition continues, we assume.

166. FDIC: *Government-sponsored sleep aid.* The Federal Deposit Insurance Corporation guarantees savings of up to $250,000 per customer per bank, and higher amounts for other types of accounts. Without it, nervous Nellies would be triggering bank runs at the first whiff of scary news. With it, even in a severe recession they can doze in peace.

165. Ingenious Baby Gear: *Lightening the burden of parenthood.* Every year 4 million babies make their entrance in the United States, reenergizing the $2.8 billion baby equipment industry. A Bumbo helps your floppy baby sit; a Snuzzler holds baby's head straight; a mirror helps you watch baby while you drive; a vibrating crib lulls baby to sleep; a stroller mosquito net filters dangerous UV rays; a microwave sterilizer sanitizes bottles. We like the handy nursing bracelet that helps you remember which side is ready for sipping.

164. George Gershwin's Music: *American originals.* Musicologists say his "Rhapsody in Blue" changed the course of American music. This 1924 jazz concerto—complete with an electrifying, opening clarinet glissando—mixed popular music rhythms with jazz-band harmonies. Inspiration supposedly came from the rhythms of the rails on a train trip from New York to Boston. "'Rhapsody,'" composer Gershwin said, "was a vivid panorama of American life." His prolific pen also created the soulful opera *Porgy and Bess* and the tone poem *An American in Paris,* which evoked the energy and sounds of the French capital, complete with Parisian taxi horns. Comfortable on Broadway as well as in classical concert halls, Gershwin wrote a slew of ever-popular standards, ranging from "I Got Rhythm" to "Swannee" and "Somebody Loves Me."

163. Prairie Towns: *Links to the pioneer past.* We salute villages struggling to survive in America's heartland after a century of declining population. Places like Marmarth, North Dakota, down to 140 people from 1,300 in the early 20th century. "Stop and see us; you'll be a stranger only once," the town Web site beckons.

162. Microsoft: *Computer colossus.* Microsoft is to computer technology what Ford was to cars a century ago. The gateway (pardon the pun) to personal computers for the masses, it grew into one of the biggest companies in America while remaining true to its entrepreneurial roots. Its cofounder, Bill Gates, who usually tops the list of the world's richest people, is a hero to nerds everywhere. A few years ago he turned his attention to philanthropy at the $39 billion Bill & Melinda Gates Foundation, which doles out more than $3 billion a year to fight poverty and illness.

161. The Anti-Smoking Movement: *Breathing free.* Studies show cigarettes can cause lung cancer not just in smokers, but also in those who breathe in their smoke. Nevertheless, nearly 20 percent of American adults continue to light up, and almost half the nonsmokers inhale secondhand smoke. Still, we've come a long  way, baby, since Berkeley, California, pioneered smoking bans in 1977: 30 states have enacted some kind of smoke-free laws, impacting more than 70 percent of the U.S. population. And the reforms continue. In 2006, Arkansas became the first state to ban smoking in cars carrying children under age six and weighing fewer than 60 pounds. Louisiana, Bangor, Maine, Rockland County, New York, and Puerto Rico have followed suit. Teens seem to be getting the message too. In 2007, 20 percent of high school students reported smoking in the previous 30 days, about half the rate 10 years earlier, reports the American Lung Association.

160. Community Colleges: *The way most people go to college.* We fictionalize the college experience as an idyllic four-year interlude for our darling 18-to-22-year-old population. But 60 percent of college students—about 10 million people annually—enroll at community colleges to fulfill their dream of higher education. Nearly 1,300 community colleges in the United States provide an affordable bridge to a university degree. These junior colleges, as they are sometimes called, provide open admission for anyone with a high school degree, and award associate degrees to students who complete the two-year curriculum. They train students in much needed technical areas, such as nurse's aide, heating and air-conditioning technician, computer repairs, and other postsecondary programs that require specific vocational skills.

159. The Two-Party System: *Divided we stand.* American politics are stable because radicals on both the left and right fail to find a large audience until they make their ideas palatable to either the Democrats or Republicans. And the two goliaths are good at stealing the good ideas of third parties. President Franklin Roosevelt called the system "one of the greatest methods of unification and of teaching people to think in common terms." And its simplicity appeals to Americans, for whom politics are a form of entertainment, like a race between two thoroughbreds. Pity the gridlocked governments in multiparty countries like Belgium. A recent unstable coalition there included five parties: Christian Democrats and Liberals who spoke Flemish; and Liberals, Socialists, and Christian Democrats who spoke French.

158. Red Sox Nation: *That says it all.* The people's team. Year after year, Red Sox fans pour into stadiums across the country, leading Major League baseball in average road attendance. They have sold out every game at Boston's Fenway Park—the oldest Major League stadium still in use—since May 15, 2003, and are a big draw for their rivals' backers too. The Sox won the first World Series, in 1903, and four more from 1912 to 1918. Then, in 1919, the "curse of the Bambino" took hold when Babe Ruth was sold to the Yankees, ending the team's dominance and giving rise to the greatest rivalry in professional American sports. The Sox finally broke the "curse," winning the Series in 2004 and 2007.

157. Home Sweet Home: *A stake in the future you can rejoist in.*
Despite the recent wave of foreclosures, more than two out of three dwellings in the U.S. are inhabited by their owners. Even George W. Bush haters confess his "ownership society" theme struck a chord. Sure, wanting an abode to call your own is a universal impulse, but it's so much easier to act on it in a country with lots of cheap land, a high standard of living, and the much beloved mortgage deduction.

156. Washington, D.C.: *Classy capital.* In 1791, President George
Washington tapped architect and civil engineer Pierre L'Enfant to design the new Federal City. The mercurial French-born L'Enfant didn't last long, but his basic plan—a grid of streets intersected by boulevards, dotted with fountains and statuary—did. Now the center of national government, the once swampy outpost is green and expansive, with no skyscrapers marring its architectural beauty. The city's unique assets inspire American and foreign visitors alike, from the Capitol, White House, and the sprawling Smithsonian Institution—with its 19 museums—to the Lincoln and Jefferson memorials and the Tidal Basin, its shores abloom with pink and white cherry blossoms every spring.

155. *Gone With the Wind*: **The best historical romance novel you can give a damn about.** More than 1.5 million copies of this 1,037-page tome by Margaret Mitchell (published in 1936) were sold in its first year. The book has continued to rank as one of our most successful best sellers. On the eve of the Civil War, Scarlett meets Rhett, loves Melanie's beau Ashley, yet marries Charles. Charles dies, Scarlett marries Frank, her sister's boyfriend. Frank dies in a Ku Klux Klan raid. Scarlet marries Rhett, but still pines for Ashley. Ashley's wife dies. Scarlet realizes she loved Rhett all along. Rhett doesn't give a damn and leaves. Tomorrow is another day. The End.

154. Alcoholics Anonymous: **Where nobody knows your name (but they care).** Bill W. and Dr. Bob started it in 1935 to help alcoholics like them recover sobriety. Now AA is a worldwide movement because its 12-step program has a high success rate. "The only requirement for membership is a desire to stop drinking," the AA liturgy says. AA's methods can help people addicted to drugs, gambling, and overeating too. For the 75th anniversary meeting in July 2010, AA selected the 65,000-seat Alamodome in Texas. San Antonio bars did not expect spillover business.

153. The Neighborhood Bar: **Where everybody knows your name (and if they don't, who cares?).** You walk in and somebody you know says, "Hey." You say "Hey" back. You decide what you will drink and how fast and whether you will stand or sit. Maybe you tell a pal about your woes or joys. Maybe you don't. You watch the TV. Or not. Nobody is in charge of you, not even you. Life is good, for now.

152. Small Town America: **Another place where everybody knows your name.** There's something comforting about keeping your front door open without fear. While most Americans live in soothing locked anonymity, we celebrate hamlets where people watch out for each other (and hope they don't gossip tirelessly behind each other's backs).

151. Hunting Wild Game: *Tracking movable feasts.* We deplore shooting animals for fun, but most of the country's 12.5 million hunters bag their prey for food, and we support them lock, stock, and barrel. They also cull surplus foragers like deer (which number 20 million today, up from 300,000 in 1900), fight to conserve woodlands and open space, and, as the backbone of a $23 billion industry—from travel to guns and other gear, to magazines to land leases—provide jobs for hundreds of thousands and give a shot in the arm to the economy.

150. The Library of Congress: *World's largest reference room.* Begun in 1800 with a $5,000 grant from Congress, this library was intended solely for lawmakers. By 1814 it housed 3,000 volumes, but all were destroyed when the British torched the new Capitol, which included the small library. Former President Thomas Jefferson came to the rescue, offering his extensive collection of 6,487 tomes, and the foundation was laid for a great national library. Today it occupies three gigantic, Italian Renaissance–style buildings on Capitol Hill and is considered the greatest library on earth. Still a congressional resource, the library also is open to the public. It boasts over 138 million items, including more than 29 million books and other print materials in 460 languages, some 58 million manuscripts, the biggest rare book collection in North America, and the world's largest collection of legal materials, films, maps, sheet music, and sound recordings. It's also home to the U.S. copyright office, which gives legal protection to the authors of original works.

149. High School Marching Bands: *The music men (and women).* When your football team stinks up the joint, you can still get a good crowd at the game if your marching band plays loudly and in tune. These moving pep bands, which began performing around 1900, combine the marches of John Philip Sousa with the tradition of military bands. They march in precision and wear uniforms and helmetlike shako hats, but have evolved to include popular music selections and dance numbers. In addition to supporting athletic programs, marching bands earn bragging rights by winning local and state competitions. We admit to a sentimental attachment to the West Orange, New Jersey, Marching Mountaineers.

148. *On the Waterfront:* *The bum also rises.* "I coulda been a contender," Marlon Brando laments as aging former boxer Terry Malloy in the often parodied taxi scene from this 1954 Oscar-winning Best Picture. The no. 1 method acting movie has Leonard Bernstein's memorable score and two messages: It's okay to rat on your pals if they are evil. Never stand under a ship's crane that's offloading Irish whiskey.

147. **Ebenezer Baptist Church:** *Rock of sages.* Founded in Atlanta in 1886 by a former slave, Ebenezer played a central role in the civil rights movement as the house of worship where Martin Luther King, Jr., preached as copastor with his father. Today it is part of the 35-acre Martin Luther King, Jr., National Historic Site.

146. **Hot Dogs:** *Top dogs at ball-parks.* Frankfurters, franks, wieners, weenies. No matter. These cooked sausages in buns come topped with mustard, onions, relish, cheese, even chili. The idea came from Frankfurt, Germany, but theirs were pork. Ours are all  beef or a blend. We put 22 million away every baseball season. So stuff your sushi, Seattle's Safeco Field. Can your Rocky Mountain oysters, Denver's Coors Field. We'll scarf a dog or two at Boston's Fenway and watch the next play.

145. **Mayo Clinic:** *Minnesota's contribution to medicine.* Like other renowned American hospitals on our list, Mayo Clinic traces its roots to an innovative 19th century physician. The modern health center in Rochester, Minnesota, with satellite facilities in Arizona and Florida, treats more than a half million patients a year.

A pioneer in heart surgery, Mayo Clinic is in the forefront of research and treatment of Parkinson's disease. It ranks first in the *U.S. News & World Report* list of best hospitals in endocrinology, neurology, and gastrointestinal disorders.

144. Fallingwater: *Wright on.* A National Historic Landmark in Bear Run, Pennsylvania, this is architect Frank Lloyd Wright's masterpiece: a reinforced concrete, sandstone, and glass home that "floats" over a 30-foot waterfall. Built in 1939, its textile-block structures and organic materials fuse it with Nature—a Wright trademark—making it an important contribution to American residential design and a major influence on aspiring architects. Some 130,000 visit each year.

143. Credit Cards: *Plastic money for the masses.* Before American Express got the ball rolling in the late 1950s, people had to carry checks and cash wherever they went. Today an estimated 180 million Americans hold 1.5 billion credit cards, giving them instant access to just about anything they want to buy. Sure, many get stuck with exorbitantly high interest rates on unpaid balances, and 30 million of them are victimized by identity fraud each year. Yet, you can beat the system by paying your bills on time and in full, gaining the equivalent of an interest-free floating loan while racking up frequent-flier miles into the bargain.

142. NAACP: *Rights champion.* The National Association for the Advancement of Colored People, the oldest civil rights organization in the country, has fought long and hard for political, educational, social, and economic equality of rights for African Americans. Although the stakes have been high, so have the dangers: firebombings, lynchings, and assassinations, especially during the racially charged 1960s. But ever since its creation in 1909, the Baltimore-based association has persevered, winning legal cases, influencing lawmakers, swaying Presidents and, ultimately, public opinion. Unquestionably, the NAACP has changed the course of American history—for the better. Especially noteworthy: its efforts leading to desegregation of public schools in the landmark 1954 Supreme Court decision, *Brown v. The Board of Education.*

141. Celebrating Our Differences: *The melting pot in reverse.*
When a fellow American asks our nationality, we usually announce our ethic roots. We're Irish, German, Jewish, British, Czech, or Indian. We celebrate Chinese New Year, Mardi Gras, St. Patrick's Day, and Greek Independence. New York City festivals, for example, include celebrations for Scots, Koreans, Haitians, Cubans, Swedes, Germans, African Americans, Italians, Polish, Tibetans, Puerto Ricans, Hispanics, Jamaicans, Jews, West Indians, Asians, Mexicans, Filipinos, and Indians.

140. High School Football: *The pride of wherever.*

Let's hear it for the Glascock County H.S. Panthers. From 1990 to 1999 the East Georgia team lost 81 straight games. It was a learning experience. The kids and coaches worked harder and eventually they started winning. But high school football isn't all about winning. It's about teamwork and winning. And sublimating individual glory and winning. It's about adolescent swagger, parents rooting for their boys, and winning. It's about old men reminiscing in taverns about the team 30 years ago that was so good it would have gotten into the regional semifinals of the state championship but for that damn interception. "For cryin' out loud. *We coulda won.*"

139. Job Hopping: *Secret economic weapon.* By age 42 the average American has changed jobs 11 times. Probably no other country has such a flexible labor market; Europeans are too busy taking five-plus weeks of vacation (and 20 days of sick leave) a year on their way to retirement at 55 to try it out. Americans, meantime, are working their tails off, jumping from one employer to another to flex their career muscles. Historically, when they have gotten laid off, nearly 90 percent have found jobs within a year, twice the European rate. Our open labor market is a boon to women, who can drop out to start families and return anytime. It also results in higher per capita output and income, higher productivity, lower unemployment, lower taxes, and fewer inner-city riots by idle youths than most nations can boast, and it upholds our sacred right to tell our bosses, "Take this job and shove it."

138. Christmas in New York: *Gotham all aglow.* Colored lights abound, ice skaters twirl on Central Park's rink, lines snake from FAO Schwarz, and chestnuts roast on vendors' fires. From Rockefeller Center's towering lighted tree to decorated windows at Lord & Taylor and Saks Fifth Avenue, it's the quintessential Yuletide celebration.

137. Barbecue Ribs: *Pass another napkin.* Though they originated in the South, the best ribs—one of our favorite meals—can be had at Arthur Bryant's Barbeque on Brooklyn Avenue in Kansas City, Missouri. The only way to eat baby back pork ribs smothered in tangy sauce is with your fingers. Dig in and remember to wipe your chin.

136. Franchising: *Entrepreneurship lite.* You own and run the venture, but pay licensing fees to a franchiser for the brand name, trademarks, marketing, and business format. It is an ideal job for people who want to run the show at minimal risk. It is a pillar of the economy, accounting for 18 million jobs at more than 900,000 franchises and an output of $1.5 trillion. In the U.S. it began in the 1850s with Singer Sewing Machines, and got a huge push in the 20th century from McDonald's, which now has 31,000 restaurants worldwide. The cost of buying a franchise doesn't have to be sky-high; total investment capital required for an All Tune and Lube auto-repair shop, for example, is about $125,000.

135. Manufacturing: *Factoring in the factories.* Like farming, manufacturing is what this country was built on, but recently this sector has been struggling. From 2000 to 2008 we lost 3.8 million manufacturing jobs—a drop of 22 percent. At the same time, imports rose 29 percent. Hardest hit have been textiles and clothing manufacturing, with many of those jobs exported. Yet, the picture isn't bleak: U.S. companies have shifted toward high-end manufacturing as the production of low-value goods moves overseas, meaning for the dollar value of the goods this country produces, it is far and away the world's top manufacturer. The value of manufacturing in the United States reached a record $1.6 trillion in 2007—double what it was 20 years earlier. We're big makers of planes, missiles, and space equipment, cars and car parts for building autos in U.S. plants, farm equipment, computer chips, and gas turbines. Still, job loss is a major worry. So, if we're going to keep manufacturing strong, Washington and Wall Street need to figure out a solution.

134. Girl Scouts: *Helping girls grow up.* Fifty million women in the United States today count themselves as Girl Scout alums. Founded by Juliette Low in 1912, the Girl Scouts once provided extracurricular fun for girls aged seven through 18 during eras when there were no other activities available. Now, girls have opportunities for after-school sports and jobs, but the Girl Scouts still attract young women to learn domestic, artistic, outdoor, and life skills. Girl Scouts are best known for their annual cookie drive, which sells more than 200 million boxes to fund activities of our 4 million Girl Scouts.

133. **The Great American West**: *Dizzying kaleidoscope of man, beast, and terrain.* It started with the Louisiana Purchase from the French in 1803 of more than 800,000 square miles of land west of the Mississippi River for three cents an acre. President Jefferson's coup doubled the size of the fledgling United States and opened the way for its expansion to the Pacific. As the frontier moved across that immense space, the players in the drama left behind an enduring national narrative of unbounded freedom and adventure. Myth or not, what players they were: cowboys, Indians, pioneers, gunslingers, posses, mountain men, trappers, fur traders, cattle rustlers, mule skinners, train robbers, riverboat gamblers, the cavalry, the Mormons, Pony Express riders, hunters, miners, ranchers, gold prospectors. The heroes and villains had names like Indian chieftain Sitting Bull, gunslinger Billy the Kid, frontiersman Daniel Boone, cowgirl Annie Oakley, and lawman Wyatt Earp. They're gone, but the prairies, the Rockies, the salt flats, the Badlands, the Grand Canyon, the bighorn sheep, the grizzlies, the buffalo, and the cactus remain. So does the solitude. So does the myth.

132. **San Francisco**: *Repository of hearts left behind.* Visitors can walk a big chunk of this city of breathtaking beauty in a single day but, alas for them, after climbing the flowered pathways to the Coit Tower, strolling through Haight-Ashbury and the Pacific Heights, wandering the Presidio park, taking a cable-car ride, visiting Fisherman's Wharf, and gazing down from the Golden Gate Bridge on the fog-shrouded bay, they have to leave. Their reluctance to go is not surprising in a city born of a gold rush. When we returned home, we were humming "I Left My Heart in San Francisco," Tony Bennett's signature song.

131. Scrabble (64 Points): *The best word game ever.* Unemployed architect Alfred Mosher Butts created this word/board game that blends skill and chance during the 1930s. He studied the *New York Times* to understand letter usage, figured out the perfect balance of consonants, vowels, and difficult letters (V is the most despicable), and gave each letter a point value. The object of Scrabble, which was first called Lexico, is to use the letter tiles you draw in a word that connects with at least one letter to words that are already played. The game has spawned international tournaments, Internet scandals (an unauthorized version gained enormous popularity on Facebook), and a Web site that publishes fake game boards to make the biggest scores ever. Given the right board positioning and triple word possibilities, the word "benzoxycamphors" can yield 1,962 points.

130. The Supreme Court: *The final authority.* The Supreme Court is an inviting target for partisans. (See *Bush* v. *Gore, Roe* v. *Wade,* and FDR's attempt to expand the court and pack it with New Deal adherents.) But the nine justices and the rest of the federal judiciary are the mechanics who keep the rule of law well oiled. A central court would be better than independent state tri-  bunals passing final judgment over the same causes, Founding Father Alexander Hamilton suggested, and he was right. The President with Senate consent picks the members, but lifetime tenure protects the justices from further politics. The court's usual glacial speed of decision-making can act as a brake on the other branches' passions. And the justices' tradition of avoiding publicity keeps at least one branch of the federal government focused on substance.

129. The Presidency: *Hercules chained.* The President has the power to order a nuclear strike and alter the world economy with an executive order. If he wants to get away, he summons *Air Force One,* and if he is in the mood for music, he can invite Stevie Wonder to the White House to jam. But all this might is usually rendered hollow by political realities and a tradition of restraint. Aides, generals, and lobbyists whisper, "Mr. President, we would advise against that," Congress and the Supreme Court can kill a proposal or a law. An intern can set impeachment proceedings in motion. Bureaucratic inertia delivers the coup de grace. President Harry Truman summed it up by describing the fate awaiting his successor, Dwight Eisenhower. "He'll be sitting there all day saying, 'Do this, do that,' and nothing will happen." The framers of the Constitution, who set it up this way because they feared an all-powerful chief executive, are high-fiving in heaven and saying, "We got that one right."

128. Congress: *The voice of the people.* The Bill of Rights keeps the government from doing evil. Doing good is the job of the 100 Senators and 435 members of the House of Representatives. They pass laws, write the budget, and help constituents petition the executive branch. Congress gets a bad rap because of a few lazy, senile, or corrupt members. The lyricist of the 1956 Broadway musical *L'il Abner* tapped into the do-nothing image when he wrote, "They sits around this place they got / This big congressional parking lot / Just sits around on their you know what / Up there they call them their thigh bones." But without Congress, America would be governed by a single-voiced chorus called the President. And members who stick around long enough can become statesmen. Like Robert Byrd of West Virginia, who in 2006 broke the record for longevity in the Senate: 48 years. In that span, he morphed from a segregationist to a civil rights champion.

127. Separation of Powers: *The nation's balancing act.* The Constitution's assignment of different duties and rights to the executive, legislative, and judicial branches has kept government working in sync for more than two centuries by preventing the concentration of power. How's that for a one-sentence civics lesson? Political scientists of all ideologies have worried lately that the balance has shifted too far to the executive branch and its massive bureaucracy.

126. Backyard Decks: *Paradise on planks.* In our imaginations, nothing bad ever happens on these peculiarly American outdoor retreats. You can loll in a hammock, doze in a chair, settle down with a newspaper and a hot cup of coffee in the morning, knock back beer with your buddies in the afternoon, and fire up the barbecue in the evening, with time-outs for admiring the hummingbirds, watching the squirrels at play, and gazing in wonder at rainbows and sunsets. It's the one place where you can say of any duty, "That can wait."

125. State University Systems: *Lifeline for the financially strapped.* Yes, state universities attract a fair share of well-off students to outstanding programs like engineering at New Jersey's Rutgers or Iowa State's CNN internship. But these state networks of higher learning are truly a blessing for striving middle- and lower-income hopefuls who can't afford a college education elsewhere. Take the California State University network, the nation's largest four-year system, with 23 campuses serving 450,000 students. For 2009–10, undergrad tuition, room, and board at California State's Northridge campus ran $15,673 for state residents. The same applicant would have been charged $48,630 at the small liberal arts Seaver College of Pepperdine University in Malibu.

124. Super Bowl Sunday: *Our de facto national holiday.* Pastoral tribes gathered annually to feast, gamble, and watch athletic contests. Super Bowl Sunday is today's global village equivalent. More than 105 million people tuned in for the 2010 game. The U.S. Department of Agriculture says it's the second biggest food consumption day after Thanksgiving. Pizza leads the junk parade at the average party. Football fans in the group get caught up in the speed, strategy, and skill of the game, whose winners get a big ring and a visit with the President. The nonfans can be confused by the rules. Why is "holding" a penalty, but four 300-pound men crashing into a slender ball carrier is not? Don't think about it. Just watch the clever commercials and the halftime show and be happy.

123. **The Catholic Church in America**: *Mass for the masses.* The biggest single profession of faith in the U.S., with 67 million members, it is hugely influential in politics and in social and theological debates. Its parishioners feed and clothe the homeless, reach out to drug addicts and women with problem pregnancies, and provide refuge for people troubled by guilt or depression. Catholic Relief Services, the U.S. arm of Caritas Internationalis, is a leading force in the fight against world poverty, disease, and natural disasters. Even unbelievers stand (or sit) in awe at the architectural majesty of Catholic houses of worship like St. Patrick's Cathedral in New York City.

122. **Sneakers**: *Fleet footings.* From rudimentary rubber-soled plimsolls in the 18th century to Keds in the 1890s, sneakers got around. (They acquired their name because they were so quiet, a wearer could sneak up on you). Soon, we had Converse All-Stars, the first shoe made just for basketball. And in 1924, German Adi Dassler created a sneaker he named after himself: Adidas. It became the most popular athletic shoe in the world and helped track star Jessie Owens win four gold medals at the 1936 Olympics. Adi's brother Rudi jumped in, establishing the famous sports shoe company Puma. By the 1950s, sneakers, originally worn mainly as sports shoes, were taking off as fashion statements for kids. In 1984, when Michael Jordan agreed to wear a Nike shoe called Air Jordans—the most famous sneaker ever—we all should have had Nike stock. Now we don sneakers for everything athletic, including walking, skateboarding, and "cross training."

121. **Honest (Relatively Speaking) Taxpayers**: *1040 reasons to love the IRS.* With so few income tax returns being audited—the rate for millionaires hovers around 6 percent—Uncle Sam relies on honesty to collect its trillions. In many other countries you don't tell the government how much you owe. The government tells you.

Americans hate taxes, but for every cheat there's probably a filer who is too rushed or lazy to claim all the deductions allowed.

120. The National Football League: *One (and millions) for all.*
It is the most successful professional sports league in the country because it has been tailored for television and the owners sublimate their egos and personal interests to the financial health of the league. The average NFL franchise is worth almost $1 billion, according to *Forbes* magazine.

119. The Right to Privacy: *Free to be antisocial.* Perhaps because they couldn't foresee how easy it would for busybodies and Big Brother to browse through our personal lives, the Constitution's framers never even thought of including an explicit right to privacy. The modern Supreme Court did the job in the late 20th century with decisions like *Griswold* v. *Connecticut*, which overturned a ban on contraceptives. At a time when government and commercial snoopers are listening to our phone calls, monitoring our Internet browsing, and using E-ZPass and credit card purchases to track our movements, we agree with 1920s and 1930s reclusive screen star Greta Garbo, who declared, "I never said I want to be alone, I said I want to be left alone."

118. Hispanic Culture: *Hablamos español!* They introduced us to potatoes, tomatoes, chocolate, margaritas, beer with lime, tacos, enchiladas, and paella. They brought us singing superstars Julio Iglesias and Ricky Martin; baseball phenoms José Reyes and Roberto Clemente, Football Hall of Fame tackle Anthony Munoz; movie stars Jennifer Lopez and Freddy Rodriguez. But wait, there's more. By 2050, the U.S. Census Bureau reports, minorities will become the majority— in large part thanks to the burgeoning Hispanic population. The nation's 47.8 million Hispanics make up 15.5 percent of the population. More importantly, Spanish has become our informal second language. In fact, the United States is the second largest country of Spanish speakers, after Mexico.

117. The U.S. Army: *Grounded in glory.* The Army's mission is to "conduct prompt and sustained combat operations on land," the field manual says. The operative word is "land." The Army's ground forces have been the largest of U.S. fighting elements from the battles of Bunker Hill to Bull Run, from Bastogne to Baghdad. GIs in combat always worry about their chances of survival. The unshaven, sleep-deprived grunts in Bill Mauldin's Pulitzer Prize–winning cartoons of World War II considered themselves "fugitives from the law of averages." Now the odds have swung in the soldiers' favor, with better medical care and technology. The modern Army, which had 569,000 soldiers on active duty at the height of the Iraq and Afghan wars, travels in Stryker land cruisers and carries rifles with infrared scopes. The service is planning a 2020 rollout of its "Future Force Warrior," equipped with enough space-age body armor, weaponry, and computers to make Darth Vader drop his light saber and raise a white flag.

116. Alaska: *Proof that size matters.* What more can you say except Alaska is really, really big: twice the size of Texas, with nearly one square mile for each of its 700,000 inhabitants, with 3 million lakes, with more coastline than the other states combined, with a 19-million-acre wildlife refuge, and with enough cold air to counterbalance the hot air in Texas. It also boasts some of the rawest beauty on earth and a plenitude of fresh salmon and halibut. Most Americans forget Russia is one of three countries that border the U.S., all because Secretary of State William Seward endured the ridicule of his compatriots in 1867 by paying two cents an acre for Alaska.

115. Diners: *Short-order wonders.* Before the neon siren songs of McDonald's and its ilk in the 1950s and later, hungry motorists pulled up to diners for cheap but solid meals like juicy charbroiled burgers with fries, or blue-plate specials like chicken pot pie and

meat loaf. Food lovers still frequent these establishments, where blowsy waitresses always call you "Hon" and you get more bang for the buck than at any fast-food chain. Take the retro South Side Soda Shop in Goshen, Indiana. Its menu features mouth-watering Philly cheese steaks, spiral fries, award-winning chili served in sundae glasses, and lemon meringue pie to diet for.

114. Bingo: *Fun from B1 to O75.* This still wildly popular game of chance emerged during the Great Depression. Attributed to New York toy salesman Edwin Lowe, players cover spots on playing cards as the letter/number combinations are called at random. The first to get five numbers in a row wins and yells "Bingo." While the rules are so simple children can play, Bingo is big business, with huge payouts in Las Vegas casinos, on the Internet, and in smoking sections of Catholic church basements.

113. Homemade Fried Chicken: *No better yum, when made by mum.* Chicken parts dredged in flour and spices and fried in fat or oil, preferably in a cast iron skillet. Few recipes are simpler than this one from the 18th century South, where even a slave could keep a chicken or two in the yard. Nowadays, chicken parts are inexpensive, so almost every family can afford a pile. The key word is "homemade." We deplore fast-food fried chicken, particularly the execrable McDonald's Chicken McNugget, which has twice as much fat per ounce as a hamburger, as Eric Schlosser wrote in *Fast Food Nation.*

112. The Internet: *The Big Bang of a new information universe.* Born in the U.S.A., with former Vice President Al Gore looking on, the Internet has exploded into the world's largest information and communications network. By our extrapolations, it will count 1.8 billion users, 200 million Web sites, and 80 billion Web pages by 2010. We are unable at this time to project the number of glazed eyes, benumbed brains, and wasted opportunities among users. Even so, it is a godsend for scholars and researchers and an instant

encyclopedia, amusement center, and communications tool for everybody. To be sure, it holds dangers (cybertheft, cyberwarfare, cyberbullying, invasion of privacy, corruption of the morals of minors), but on balance, it is a boon to humankind—closing out the 20th century with the most important invention since Orville Wright opened it with the first successful powered flight.

III. American Humor: A *potpourri of laughs*.

Ha! Ha! Ha!

From the drolleries of Mark Twain, the founding father of modern American humor ("Always do right. This will gratify some people and astonish the rest") to the dark cynicism of novelist Ambrose Bierce, to the laugh-out-loud cartoons of James Thurber, to the slapstick on-screen antics of the Three Stooges and Lucille Ball, to the goofy laugh of actor Eddie Murphy and the smirk of late night TV show host Johnny Carson, to the comedy sketches of *Saturday Night Live,* to the ironies of Dave Chappelle, to the one-liners of stand-up comedians like Henny Youngman ("Take my wife . . . please!"), to . . . well, *Mad* magazine, Scott Adams, Bill Cosby, Jerry Seinfeld, The Simpsons, Jon Stewart, Tina Fey, Demetri Martin, and a thousand ticklers of the national funny bone, American humor borrows from the resources of all the world's races and nationalities and creeds to produce the funniest stage in the world.

110. Yellowstone National Park: *Hot destination.*

Located largely in Wyoming, with portions in Montana and Idaho, Yellowstone is noted for its geysers, waterfalls, boiling hot springs, and herds of wildlife, including grizzly bears, wolves, bison, and elk. The park sits on a huge collapsed crater formed 600,000 years ago during a cataclysmic explosion. The preservation of this area in 1872 by Congress as the world's first national park did more than shield its 2 million-plus acres from private development. It established a role for the national government in protecting and administering precious wilderness areas for all people—not just the wealthy—to enjoy. Every year, tourists flock to Old Faithful, an extraordinary geyser whose regular eruptions can send scalding water as high as 180 feet in the air. In fact, a major concern for environmentalists these days is protecting the park against a tide of 3 million annual visitors.

109. **Our Single National Market**: *Engine of growth.* Equal access to 320 million consumers on half a continent has long given American industry a potent economic edge over the world's fractious and fractured regions. In the 20th century, countries like China, India, and the old Soviet Union had more people than the U.S. but were too poor and poorly run to challenge us. They someday might, by imitating us. Europe has unified, created a common currency and torn down trade barriers. China has ditched communism for capitalism. Even so, we think it will be a while before either knocks our $15 trillion economy off its perch as no. 1.

108. **Oreo**: *King of cookies.* More than 490 billion of these chocolate sandwich cookies with cream in the middle have been sold since they were introduced in 1912. Nobody, not even Nabisco, is quite sure where the name originated. Some claim it was a takeoff on the French word for gold. Others say it comes from the Greek for mountain, *oreo.* Still others suggest it was just a catchy made-up word. It's the no. 1 selling cookie in the United States. Just about everybody has a preferred method of culinary attack. Some dunk 'em, others twist off the cookie tops, and not a few lick the cream off and throw away the chocolate wafers. Whatever.

107. The Pacific Coast: *Surf's up.* From Border Field State Park on the Mexican frontier to Cape Flattery in Washington, the 1,359-mile Pacific Coast has eye-pleasing beaches and cliffs. And sunsets, except when clouds snuggle along the breakers off Washington, Oregon, and northern California. "Ocian in view, O! the joy," explorer William Clark wrote in his diary on November 5, 1805, after he and Meriwether Lewis reached the mouth of the Columbia River in their quest to be the first Americans to cross the continent. Two centuries later the coast is more than a place. It is a fantasy with music by the Beach Boys and the Mamas & the Papas playing in the background and where the normative pastimes are surfing, being a *Baywatch* babe, and drinking designer coffee in a spa facing the water in Carmel, California, Eureka, Oregon, or at Alki Beach, Washington. Or Big Sur, California, where novelists can find the peace they need to write about places of substance, like Paris and Brooklyn.

106. The Atlantic Coast: *The beckoning shore.* Historians aren't sure which European explorer first saw the eastern shore of what is now the United States. It was a desolate wilderness, soon to become the beach head for the European invasion that ended 3,000 miles later at the Pacific Ocean. Now, the 2,069 miles along the Atlantic are magnets for American migration from the interior, especially in the summer when the beaches glimmer in the sun. But it's more than just a place to bathe in seaside resorts such as Myrtle Beach, South Carolina, Ocean City, Maryland, and Virginia Beach, Virginia. It is the site of a diverse collection of cities: tony Bar Harbor, Maine, cosmopolitan Boston, New York, Charleston, and Miami, gambling haven Atlantic City and history-rich St. Augustine, the oldest continuously settled town in America, going back to 1565.

105. The Special Relationship with Britain: *Blood thickener.* Since that bit of unpleasantness known as the War of 1812, we have stood together in war and peace. How could it be otherwise? The Founding Fathers were British subjects. From the Mother Country we inherited our language, legal and political systems, and our sense of fair play. Americans love British eccentricity and the British Royal Family. Mark Twain, speaking when both countries were in warrior mode, said, "We have always been kin: kin in blood, kin in religion, kin in representative government, kin in ideals, kin in just and lofty purposes; and now we are kin in sin." British Prime Minister Winston Churchill, who led his country through World War II (and whose mother was American) coined the term "special relationship," and politicians have invoked it ever since.

104. *The Adventures of Huckleberry Finn*: Our best book. To Ernest Hemingway, all modern American literature flowed from this one novel. Its theme, setting, and southern vernacular make it authentically American, divorced from British and European culture. Author Mark Twain's 1884 masterpiece chronicles the saga of Huck Finn, a white teenager fleeing his abusive father by rafting down the Mississippi River. Huck meets Jim, a runaway slave, and they join forces. Despite society's racist drumbeat, Huck comes to value Jim as a friend. The novel may be packed with hijinks and humor, but it's really about freedom and morality. In it, Twain was blasting racism, with its concomitant segregation and lynchings. Certain southern states objected and banned his book, yet some modern readers deem it racist because of its gritty regional language.

103. Comeback Cities: You can go home again. The last quarter of the 20th century saw city residents fleeing. Crumbling infrastructures, coupled with rising crime and a prevailing sense of hopelessness, were mostly to blame. Then people got mad as hell and decided not to take it anymore. New York City, now with nearly a million more people than in 1990, cleaned up midtown and restored order; Baltimore transformed its waterfront into a tourist destination; Cleveland centered its revival around new baseball and football stadiums and the Rock and Roll Hall of Fame. This urban renewal isn't limited to the biggest cities. Special note goes to Hoboken, which morphed from a washed-up New Jersey port town into a chic spot for young professionals who hop the train each day into New York City.

102. Free Ice Water in Restaurants: *A cool dining-out perk.* In most countries, you have to request a glass of tap water and then put up with the puzzlement of the waiter, who will reluctantly serve the beverage at room temperature. In the U.S. it comes chilled, no questions asked.

101. Medal of Honor: *For none but the brave.* Of all the awards America bestows, none is more revered than the medal for conspicuous gallantry in battle. It keeps its luster because so few are given: one for every 38,000 soldiers called to arms in wars since 1917. In the presence of a Medal of Honor recipient, no matter how low a private, the highest general must salute. Winners like Alvin York in World War I and Audie Murphy and Ira Hayes in World War II became celebrities and the subject of Hollywood movies. Others, equally heroic, returned to obscurity except among their awed neighbors. Most did not return at all. Fifty-seven percent of the 978 medals awarded from the beginning of World War I through 2009 were given posthumously. Like the one to Navy Lieutenant Michael Murphy, who in June 2005 braved the bullets of 30 enemy soldiers in Afghanistan to save his outnumbered command.

100. The Everglades: *Ecological pearl in peril.* A 4,000-square-mile ecosystem in southern Florida of sawgrass marshes, cypress swamps, and mangrove forests, the Everglades have shrunk to half their original size from draining, sugar farming, and overdevelopment. They remain a national treasure, teeming with wildlife, including endangered species like the Florida panther, the American crocodile, and the wood stork, and they draw millions of tourists a year. Today, a $12 billion (the latest figure as of December '09) 30-year restoration plan, the largest environmental rescue project in U.S. history, aims to revitalize what Native Americans called "the grassy waters."

99. Special Olympics: *Where everyone is a winner.* If you have never witnessed the magic of a Special Olympics event, put it on your to-do list. It will be the best feel-good thing you do all year. The first Special Olympics began in Chicago in 1968, an outgrowth of the efforts by the late Eunice Kennedy Shriver, whose sister was developmentally challenged, to bring athletics to these special people. Now, more than 40 years later, the Special Olympics is an international movement that has done just that. It has brought sports competition and physical fitness training to the developmentally delayed. But even more notable, the Special Olympics has taught us so-called normal folks volumes about human dignity, sportsmanship, fair play, the real meaning of sports, and the lesson that if you play the game the right way, everybody wins, all the time.

98. Elvis Music: *Cultural earthquake that still gets us all shook up.* Though not the founder of rock 'n' roll, Elvis Presley transformed it into the central force in American music in the 1950s through his genius for melding elements of gospel, country, jazz, and black rhythm-and-blues. Oh yes, there was also his charisma and talent. Every pop star and music lover owes a debt to the King.

97. Clam Chowder: *Sea-scented soother.* In the 1620s, Native Americans encouraged Pilgrims to eat local fare, but they recoiled at slurping down clams, instead feeding them to their hogs. (A shame, since this broth teems with the abundant bivalves so easy to dig up along the Massachusetts shore.) That fussy mind-set meant chowder (the name may come from cauldron, a big cooking pot) didn't take hold in New England for more than two centuries.

Clam chowder's incarnation in the 1850s was thickened with cream and flavored by salt-pork-fried onions and potatoes. In the 1930s blasphemers began substituting tomatoes for cream, then tossing in bacon, green pepper, and celery. They called that Manhattan clam chowder. New Englanders were aghast. In 1939, Maine legislators introduced a bill making it illegal to add tomatoes to this chowder. Even today it's rare to find both versions on the same menu. And please, it's pronounced chow-dah.

96. Growing Up to Be President: *Equal opportunity ambition (sort of)*. We tell our children to dream big, the sky's the limit. When we ask them what they want to be when they grow up, lots of kids say President of the United States. And, of all the Americans who have ever lived, that dream came true for 44 men. The 2008 election proved that you don't have to be an old white guy to be President. The dream is alive for minority men. Women—whose hopes were dashed by President Obama's defeat of challenger Hillary Clinton—have reason to believe their turn will come someday soon.

95. Government Regulation of Capitalism Gone Wild: *Corralling the bulls.* The latest recession reminded the country why the Securities and Exchange Commission, the Federal Reserve, and other watchdogs need sharp teeth to keep out-of-control financiers from harming the economy.

94. Oil, Natural Gas, and Nuclear Power: *Take that, Arab oil sheikhs!* Despite the decline in our petroleum output since 1970, we continue to produce a lot of the stuff, about 5 million barrels a day, and some experts predict an increase. Better yet, in the past few years huge fields of natural gas have been discovered in shale formations in several states, including wells in Louisiana that increased the nation's estimated reserves by up to 65 percent. Finally, a new generation of nuclear reactors is on the horizon to supplant the ones built decades ago before cost overruns and accidents at Three Mile Island and Chernobyl put an end to construction. We're on an energy roll, baby!

93. The Statue of Liberty: *World symbol of freedom.*

The most recognizable statue on earth greets those sailing into New York Harbor. A century or more ago the Lady of the Harbor, her outstretched arm holding aloft the torch of enlightenment, was the first glimpse yearning immigrants got of America, their promised land. Her clarion call: "Give me your tired, your poor, your huddled masses yearning to breathe free," wrote poet Emma Lazarus. The copper-clad sculpture was erected in 1886, a gift from the French, who admired the young Republic's democratic ideals. Standing 151 feet tall on a 150-foot pedestal, she matched the expansiveness of the prospering country. These days Liberty is a must-see for millions of tourists. Her image graces everything from T-shirts and pillows to plates and cocktail napkins. But commercialization will never cheapen her. As historian Daniel J. Boorstin put it, "I become an American all over again when I see the Statue."

92. America's Commitment to World Peace: *Can't we all play nice in the sandbox?*

We've been trying to solve the Arab-Israeli conflicts for years. We are often the main player in United Nations peacekeeping missions. Our role as leader of the Free World puts us in a unique position to critique injustice and bully offenders by withholding aid, boycotting trade, and publicly shaming them by holding them up to world ridicule. In recent years our military peacekeeping efforts have taken us to such far-flung places as Kosovo, Bosnia, Haiti, and Somalia. When we can't keep peace, we "enforce" peace, or "build" peace, through maneuvers that often put American soldiers in harm's way. Though sometimes misguided, our commitment to world peace and our willingness to put our muscle behind our message is one of the things that makes us the people we are in the eyes of the world.

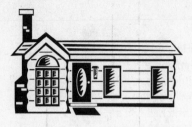

91. Habitat for Humanity:
Shelter from the storm. The organization that devotes volunteers' money and labor to home construction for the poor had its roots in the Christian service movement. The official history says, "What the poor need is not charity but capital, not caseworkers but coworkers, and what the rich need is a wise, honorable, and just way of divesting themselves of their overabundance." Since the mid-1970s, Habitat for Humanity has built homes for 1.5 million people in more than 3,000 communities around the world.

90. Las Vegas: *What happens there, stays there.* About 39 million visitors flock to Sin City annually, but only five percent admit they come to gamble at one of the 1,701 licensed gaming establishments. Some come for a quickie wedding at a casino chapel (about 115,000 marriages occur there annually). Others are lured by the 15,000 miles of neon tubing that light its famous strip and themed hotel/casinos that scream, "Disneyland for adults." For a while Las Vegas tried to shed its naughty image and advertised itself as a family destination, what with the New York, New York roller coaster, Paris's Eiffel Tower, the Venezia's Grand Canal, and the Bellagio's sound and light show. They even highlighted family side trips to the Hoover Dam, just 35 miles down the road. But, what the heck? Leave the kids home: those topless showgirls are too hard to explain to a 10-year-old.

89. Peanut Butter and Jelly Sandwiches:
The fussy eater's favorite lunch. They're inexpensive and as close as you can get to eating sweets without doing so. Some kids have them for lunch almost every school day. For the true American experience, only fluffy Wonder Bread will do.

88. Casual Dress: *The jeans in our genes.* The old sartorial standards of formal wear like sport jackets and ties for men and pantsuits and high heels for women fell first in shops and restaurants, then in trains and planes, then in offices (beginning with dress-down Fridays), and finally in houses of worship. The trend saves us time, money, and discomfort, though after seeing a funeralgoer in a black T-shirt recently, we decided at least one occasion ought to be immune to it.

87. The Fourth Amendment: *Your home is your castle—mostly.* The Constitution's Fourth Amendment guards against unreasonable searches and seizures. The Founders ratified it in reaction to English abuse of the writ of assistance, a kind of general search warrant that gave Redcoats the green light to invade private property to unearth traitorous writings or smuggled goods. Under the amendment, police search and arrest warrants must be court-sanctioned and supported by probable cause. There are numerous exceptions, such as lawful searches of wide-open spaces or seizures of contraband in plain sight. But the Supreme Court's "exclusionary rule" remains key, meaning evidence obtained through an unreasonable search or seizure can't be used against you at trial.

86. Service with a Smile: *Something to smile about.* Go into any restaurant or shop in the U.S., and someone on the staff will smile at you. Not so abroad. "Nobody ever says, 'Have a nice day' here," gripes an American friend who has lived in Spain for 40 years. We hear the same complaint from expatriates all over the world.

85. Freedom of Assembly: *Coming together for a cause.* When protesting students gathered in Tiananmen Square in China in 1989 and refused to disburse, tanks rolled in. God bless America, where those militaristic excesses (we're reminded specifically of student clashes with National Guard troops during Vietnam War protests in

1970) are uniformly condemned. The modern-day Supreme Court has consistently protected the right to peaceably assemble even for repugnant groups like the Ku Klux Klan. The civil rights movement owes much of its success to the peaceful protests protected by this First Amendment freedom. More than 250,000 protesters assembled in Washington in 1963 to witness the pivotal "I Have a Dream" speech given by Martin Luther, Jr., from the steps of the Lincoln Memorial.

84. Having Canada as a Neighbor: *The peaceful border.* Canada is often dismissed as a tediously huge and cold place filled with boring people. Other countries should have such neighbors. Think Pakistan-India, Israel-Lebanon, Iraq-Iran. Though security on the U.S. side increased after 9/11, the 5,525-mile border with Canada is the longest undefended frontier in the world, and the $600 billion a year in U.S. trade with Canada is almost as big as American commerce with China and Japan combined. We would write more, but we have run out of interesting things to say about Canada. Oh, wait. We really like pop singer Avril Lavigne and *Jeopardy!* quizmaster Alex Trebek.

83. Google: *Omnipotent search engine.* As a verb, *google* means to look up a reference on the Internet. As a noun, it's the highly profitable brainchild of Stanford University doctoral students Larry Page and Sergey Brin. Since its debut in 1998, Google gained a reputation as the fastest, easiest, and best way to track down information on the Internet. In 2008, Google muscled out all competitors with 85 billion searches.

82. The Fourth of July: *Feting the Founding Fathers.* From the boom of cannon, the rattle of artillery, the clanging of bells, and the music of a Hessian band in Philadelphia in 1777 on the first anniversary of the signing of the Declaration of Independence, to the clamor

of parades, cheering crowds, joyously off-key patriotic bellowing, and the explosions of fireworks across modern America, our most hallowed secular celebration is guaranteed to bring lumps to the throats of even hardened cynics.

81. Civilian Control of the Military: *Leashing the dogs of war.*
The tradition of presidential supremacy over the armed forces is spelled out in the Constitution, taught at West Point and Annapolis, and ingrained in our bones. Only Congress can challenge the commander-in-chief's war-making authority, notably by the power of the purse. General Douglas MacArthur was the exception who proved the rule—he openly defied President "Give 'em hell Harry" Truman during the Korean War in 1951 and was quickly shot down, despite his popularity. In his farewell address to Congress, MacArthur declared that "old soldiers never die, they just fade away." So do generals' outsized ambitions.

80. The Great Lakes: *Maritime marvels.*
Nearly one-fifth of the world's fresh surface water flows through these five giants, so vital to both the environment and commerce. The lakes—Huron, Ontario, Michigan, Erie, and Superior—separate the U.S.-Canadian borders along eight midwestern states. Their waters connect to the Atlantic through the St. Lawrence Seaway's system of locks and canals, allowing oceangoing vessels direct access to America's industrial heartland. Iron ore and grain are two of the biggest commodities shipped on these waters. Tourism and recreation loom large too. Since the 1970s, pollution prevention and control programs, along with pesticide bans and conservation plans, have scored dramatic improvements in the quality of the Detroit River and western Lake Erie. Bald eagles and peregrine falcons are back. The walleye population is soaring. Lake sturgeon and whitefish, long absent, are spawning there again. Problems persist, however, in runoff, habitat loss through invasive species, and release of toxic substances.

79. Massachusetts: *Birthplace of freedom.* The Pilgrims and Puritans set up colonies here in the early 17th century in search of religious freedom. Agitation in Boston against the British in the 18th century ended in the American Revolution and independence. In the 19th century, Massachusetts became the first state to abolish slavery and led the way in the Industrial Revolution. Notable sons and daughters include Revolutionary heroes Paul Revere and Crispus Attucks, the Adams and Kennedy political dynasties, TV journalist Barbara Walters, comedian Jay Leno, business legend Jack Welch, football great Doug Flutie, and actor Ben Affleck.

78. Disney World and Disneyland: *Mickey Mouse's playgrounds.* For more than 50 years Walt Disney's vision has defined the perfect family vacation. Disneyland was first, opening in Anaheim, California, in 1955. Then came Disney's takeover of central Florida, where Disney World and its offshoots—Epcot, Animal Kingdom, Hollywood Studios, and several water parks—were built on cattle pastures. Today, Disney World has usurped all competition. It is the largest single-site U.S. employer, with 62,000 employees. Not only does it lure more visitors annually (25-plus million), it boasts that more than 10.5 million good bugs were released around the campus last year to control plant pests.

77. Uncle Sam: *Everyone's favorite uncle.* Hooray for the Red, White, and Blue! Hooray for the well-known image of Uncle Sam! He's the scrawny white-haired eminence grise decked out in a clownish patriotic costume and an oversized top hat. Literary references to a real Uncle Sam first cropped up during the War of 1812, but the cartoonish character we know today was the brainchild of 19th century illustrator Thomas Nast. Uncle Sam has been used by the federal government to recruit soldiers and sell stamps and savings bonds. No Fourth of July parade would be complete without the requisite inclusion of the larger-than-life Uncle Sam careening down Main Street on stilts.

76. Declaration of Independence: *Words to live by.* We hold this truth to be self-evident: the 1776 Declaration of Independence is our most treasured document. Not only did the declaration penned by Thomas Jefferson sever our dependence on England, it articulated lofty ideals: the "right of the people" to determine their government; our "unalienable rights" to "life, liberty, and the pursuit of happiness." It took us almost two centuries to wrestle with the "all men are created equal" tenet, but we eventually came around. Although the Declaration of Independence was adopted by the Continental Congress on July 4, 1776, it wasn't signed until August 2, when patriot John Hancock put his John Hancock on the parchment first. You can view the original at the National Archives in Washington, D.C.

75. Social Security: *Enjoy it while it lasts.* President Franklin D. Roosevelt started it in the Great Depression and made payroll deductions mandatory to keep politicians from monkeying with it. Too bad the trust fund the country built for 70 years will run out of money within three decades at current rates of expenditure. In the meantime, the 50 million people receiving benefits will cash their checks and let the future take care of itself. President George W. Bush tinkered with the idea of limited private investment. Didn't hear much about that even before Wall Street tanked in 2008.

74. NASCAR: *Vroom with a view.* During the 1920s Prohibition era in Georgia, bootleggers hoping to escape the law would speed moonshine to market over curving, country roads in souped-up cars. When these whiskey trippers started competing against each

other for fun in their everyday cars, stock-car racing was born. The National Association for Stock Car Auto Racing, or NASCAR, was formed in 1947. Originally a good ol' boys' pastime with deep roots in the Southeast, NASCAR has gone national. It's our fastest-growing spectator sport, with races annually drawing close to 75 million people in person or via television. From the boxy cars of the 1940s to the sleek racers of the present, the risk has always been high, with speeds these days reaching 190 mph. Many of the best U.S. auto racers have driven in NASCAR events, guys like Junior Johnson, Tony Stewart, and Dale Earnhardt, Jr.

73. The Sixth Amendment: *Right to counsel.* Lawyers finish low in public rankings of professionals but they are indispensable at times, and the Constitution gives us the right to have one when hauled before the law. Great film portrayals dramatize the right in action. We see Henry Fonda as *Young Abe Lincoln* winning an acquittal with wit and humor. Spencer Tracy as Clarence Darrow defending an evolutionist in *Inherit the Wind,* Gregory Peck as Atticus Finch in *To Kill a Mockingbird,* doing his best in a lost cause, and Michelle Pfeiffer in *I Am Sam,* battling for a father's right to keep his child. The right to counsel is not a gimmick to help murderers go free or fakers with neck braces to fool juries out of $1 million. The Sixth Amendment is in the Constitution to make sure Americans in a legal contest have a champion to make the adversary stick to the rules.

72. Wall Street: *Stomping ground for bulls and bears.* The heartland of American capitalism in lower Manhattan, Wall Street is also a catchall phrase for trading in stocks, bonds, and other financial instruments. Though it came under intense regulatory scrutiny in the 2008-09 market

meltdown, and though ordinary folk bridle at the riches amassed by its masters, we love Wall Street because our economy couldn't function without it. Besides, it's fun to track the ups and downs of the Dow and other market indexes—and the gyrations in the value of our 401(k)s, 529 plans, and other nest eggs.

71. Public Beaches: *Where the people can play.*

With 12,500 miles of coastline, and thousands more on lakes and ponds, we consider access to the water our God-given right. This is supported by the Public Trust Doctrine, an age-old principle bolstered by a U.S. Supreme Court opinion in 1892. The doctrine recognizes that certain resources such as the ocean, navigable rivers, and their shores, are held in trust by the government for public use and enjoyment. Fortunately, most states allow access to the area between tide marks for walking, swimming, fishing, and sunbathing. However, some landowners still try to throw their weight around, erecting bogus "Private Property" signs or intimidating gates. Just let them try to bar us from beautiful beachfronts like Ocracoke Island, North Carolina, Smith's Point, New York, or Lowdermilk Park Beach, Naples, Florida. And sun hats off to the Los Angeles Urban Rangers for defeating territorial property owners by issuing a map showing every hidden yet public pathway down to Malibu's beaches.

70. The Greening of America: *Our gift to Mother Earth.*

We're all environmentalists now, or claim to be. To protect the planet from the ravages of years past, we have enacted laws from the Clean Air Acts of 1970 and 1990 and the Endangered Species Act of 1973 to local recycling ordinances. Nonprofit groups devoted to preserving wilderness, protecting our rivers and open spaces, reducing our carbon output, and otherwise safeguarding the ecosphere have proliferated. Movements like organic farming and alternative energy have captured the public imagination. Naysayers abound, accusing environmentalists of alarmism and elitism, but their arguments fall flat. These days, even coal companies and property developers wrap themselves in the green banner.

69. Public Education: *Taxpayers' gift to the next generation.* Since the 1830s when the idea of universal free education caught on, the number of students in U.S. public schools K-12 has grown to 53 million—nine times private school enrollment. Public schools have no ethnic, class, or religious boundaries, so they teach the art of getting along with people who are different. Or bigger. Behind successful public school alums are teachers like B. J. Frazer at Dixon H.S. in Illinois, who hooked Ronald Reagan on acting; and Kate Deadrich, who gave future President Lyndon Johnson reading lessons in a one-room school in Texas. Americans are always concerned that their public schools aren't good enough, spawning initiatives like President George W. Bush's No Child Left Behind Act. Why the fuss? A couple of the authors of this book is public school grads and were teached good.

68. Corn on the Cob: *Dream food that dribbles.* Sweet corn hasn't caught on outside the Western Hemisphere, but in the United States you can feast on its derivatives—corn stews, corn chowders, corn bread, corn custards, corn puddings, corn grits, and creamed corn—year-round. Ah, but when the crop ripens in

the summer, it is time for corn on the cob, roasted, grilled, steamed, or boiled, rolled in melting butter that will drip down your chin and sprinkled with salt and pepper. Ambrosia and nectar may have sustained the Olympian gods, but our bet is that Zeus would have reached for corn on the cob first.

67. Family Farms: Still plowing on. In
the mid-20th century, impersonal, multi-million-dollar agribusinesses sprouted on the American landscape. They seemed to sound the death knell for family-run farms where work was back-breaking but life was enriching and everybody shared

rock-solid values. Happily, that hasn't happened. After decades of decline, the number of family farms has increased by about four percent, reports the U.S. Department of Agriculture. And it's still true that the vast majority of America's 2.2 million farms—nearly 96 percent in 2007—continued to be family run. This all goes to show how we treasure our pastoral roots.

66. Our Greatest Universities: Ivory towers mostly covered
with Ivy. When we talk about America's greatest colleges, the eight Ivy League schools especially come to mind: Yale, Dartmouth, Brown, Cornell, the University of Pennsylvania, Princeton, Columbia, and Harvard, the first and still the finest. With nearly 2 million new students heading to American higher education annually, these universities pluck the very best, admitting only about 10 percent of applicants. Sometimes it matters whom you know—the surest way to assure a seat in the freshman class is when the amphitheater is named for grandpa. But connections aside, the Ivies are not the only truly great American universities. Wanna be a rocket scientist? The Massachusetts Institute of Technology tops the list. The pedigree of a Stanford or a University of California at Berkeley diploma also opens doors.

65. Big Breakfasts: Going whole hog. Fruit or juice. Bacon, ham,
sausage. Hell, why not all three? Eggs, pancakes, French toast, grits

or potatoes. Biscuits, toast, a muffin or bagel. A sweet bun or Danish and a couple of mugs of coffee. Who needs those arteries, anyway?

64. The Dollar: *The buck that stops nowhere.* Its glory days might be over—rarely do you hear the phrase "sound as a dollar" anymore—but the greenback remains the world's reserve currency and safe haven in times of war and economic upheaval. The poor of the earth hoard it. Criminals covet it for its portability. Americans love the crumpled notes that depict Washington's weary visage, Lincoln's stern gaze, Hamilton's knowing glance, and Jackson's glare for their boring familiarity.

63. Volunteerism: *Deliberate acts of kindness.* Following September 11, President George W. Bush urged distraught Americans to channel their angst into volunteering in their communities. The request was not necessary. Americans have always had a strong tradition of volunteerism, whether in government, schools, churches, hospitals, or nursing homes. Many simply want to help out and give back. Some also crave social ties and a greater sense of purpose to their lives. Whatever the motivation, this willingness to work for others without financial gain has filled the ranks of organizations like the Peace Corps, Habitat for Humanity, Rotary International, the Red Cross, and the League of Women Voters, to name a few. More than 60 million Americans volunteered their services between September 2006 and September 2007, and spent a median of 52 hours doing so.

62. Mutts: *Dogs for a melting pot nation.* Forty percent of U.S. households have at least one dog, and the five top breeds from year to year are Labrador retrievers, Yorkshire terriers, German shepherds, golden retrievers, and beagles. Put them all together and you get a true American dog—a mongrel. Mutts are as

diverse as the country, ragged around the edges perhaps, but loved for who and what they are, not because they live up to a kennel club definition. The *American Heritage Dictionary* says the term "mutt" came into use as a shortened version of the word "muttonhead," or stupid person, and was first applied to undistinguished dogs in the early 20th century.

61. Welcome Mat for Refugees: *Opening the Golden Door.* Our record is far from perfect—turning back 915 Jewish refugees from Adolph Hitler's Nazi Germany who had sailed on the SS *St. Louis* from Hamburg to Florida in 1939 was shameful—but even so, the hospitality we show to the world's oppressed is unrivaled. From Hungarians fleeing Soviet tanks, Cubans escaping Fidel Castro's island prison, and Vietnamese on the run from Communist dictatorships, to the more recent exodus of victims of political persecution in Bosnia, Burma, Somalia, Sudan, and Iraq, we invited the "huddled masses yearning to be free" of Emma Lazarus's 1883 poem to join America's melting pot.

60. The National Guard: *Citizen soldiers.* They go back 400 years to English reliance on local militia to train together in defense of the colonies. These days, the Guard answers to its state commanders-in-chief, the governors, unless mobilized for national service—an arrangement created by the Founders to avoid a federal monopoly on military power. We applaud the Guard's efforts in the Gulf States following Hurricane Katrina in 2005, when nearly 60,000 members answered the call, the largest and fastest deployment of U.S. forces ever for a domestic crisis. They helped rescue more than 17,000 and evacuated 70,000 whose homes were destroyed or damaged. Meantime, Guard participation in the War on Terror continues. Since September 11, more than 200,000 Guard soldiers have been mobilized for active duty overseas.

59. Free Press: *Democracy's catalyst.* The bright light among the Founding Fathers, Thomas Jefferson, explained the importance of a free press: ". . . [Were] it left to me to decide whether we should have a government without newspapers or newspapers without a government, I should not hesitate a moment to prefer the latter." The concept of a free press gets passing mention in the First Amendment, and journalists have cloaked themselves in that protection ever since. There have been temporary setbacks. Most notably was President Nixon's audacious attempt to stifle the 1971 publication of the Pentagon Papers, a top secret report highly critical of the U.S. involvement in Vietnam. Kudos to the Supreme Court, which had the wisdom to uphold Jefferson's vision.

58. Our 24/7 Business Culture: *All retail, all the time.* What's up with those so-called civilized countries like Italy and Spain where the stores close after lunch and the restaurants are open only at traditional mealtimes? America is the land of the all-night diner, the 24-hour bank, the gas station that never sleeps, and the fast-food drive-in open 'round the clock. Can't sleep at three A.M.? Do your weekly food shopping at the Winn-Dixie or Safeway. Keeping the jaws of commerce munching at all times feeds our nonstop creativity and provides a lively market in second jobs. The down side: armed robbers love 24-hour pharmacies, as drug chains in Florida discovered in 2008.

57. Our Judeo-Christian Culture: *Infrastructure of our greatness.* Pipe down, you postmodern multiculturalists. It is a simple statement of facts: most Americans are Christians, a religion born of Judaism, and millions are Jews. They share a common heritage and set of beliefs, notably love of God, love of neighbor, a thirst for justice, and recognition of the human capacity for wickedness. These God-given principles, inscribed in our hearts and in our laws, have enabled our country to flourish as a free and independent democracy. End of sermon.

56. Redwoods: *Seeing is disbelieving.* The tallest trees ever recorded, these cousins to the Giant Sequoias of the Sierra Nevada region are at their highest along the northern California coast. Old-growth redwoods can exceed 300 feet in height and 18 feet in diameter. Some go back 2,000 years. Motorists can ogle these towering patriarchs along the 33-mile Avenue of the Giants in scenic Humboldt Redwoods  State Park. Hikers can trek through Redwood National Park in Orick, site of the largest virgin groves of ancient redwoods on earth. The park's Tall Trees Grove boasts the world's loftiest redwood at 366 feet. If all that weren't enough, the national park's 113,200 acres include 40 miles of dazzling Pacific coastline.

55. American Fashion: *Anything goes.* We researched great American fashion designers like Calvin Klein, Michael Kors, Ralph Lauren, Donna Karan, and Tommy Hilfiger, but it basically boils down to this: America gave the world T-shirts, sneakers, and blue jeans. Everything else is derivative. We celebrate Americans, however, for their devil-may-care, what-are-you-staring-at?, in-your-face outfits that we call individual fashion statements.

54. Basketball: *Great fun for all those tall folks.* The average height of a professional player in the National Basketball Association is six-foot-seven (although retired star Muggsy Bogues was only five-foot-three). Yet another uniquely American sport, basketball has grown from its birth in 1891 into an international phenomenon and a popular Olympic sport. Basketball is a cheap, simple sport that attracts about 12 million children to the court annually.

Its appeal is obvious: you can play alone and all you need is a ball, a hoop, and the dream of exceeding the exploits of the great one, Michael Jordan.

53. Being the Beacon of Liberty to the World: *An idea worth pursuing.* Puritan minister John Winthrop preached in 1630 that the new Massachusetts Bay Colony would be "a city upon a hill; the eyes of all people are upon us." President Reagan quoted those lines to tell Americans the world looked to them as the keeper of the flame of freedom. Acting as if we have a natural right to guide the world is one of the worst things about America. Conducting ourselves so we deserve to be the guide is one of the best. Not for nothing did anti–Vietnam War demonstrators clashing with the police in Chicago in 1968 chant, "The whole world is watching." When America gets it right, it can be what President Abraham Lincoln called "the last best hope of earth."

52. Poker: *America in miniature.* Every game is a new beginning, every shuffle of the deck a second chance. The past doesn't matter, and you can win big with the right balance of recklessness and calculation. The French brought *poque* to New Orleans, and Mississippi riverboats carried it into the heartland. It became a mark of American manhood and a passion of the likes of pioneer Jim Bowie and statesmen Henry Clay and Daniel Webster in the 19th century; Harry S Truman and Richard Nixon in the 20th, and Amarillo "Slim" Preston, Jr., well into the 21st. Hawaii's 19th century "Merrie Monarch" King David Kalakaua, in flight from a mob, paused by the roadside to finish a hand. Fifty million Americans play regularly, though few can match Herbert O. Yardley's tale of the game's fickleness in *Education of a Poker Player* of how a man bet his corn farm against a tent show, "only to die three minutes later, his cards clutched in his hand—a winner." Interesting. Now shut up and deal.

51. The Interstate Highway System:

People on the move. As U.S. forces raced along autobahns in pursuit of Hitler's retreating army, Supreme Allied Commander Dwight Eisenhower marveled at the Germans' sleek, four-lane divided highways with on- and off-ramps—and no traffic lights. As President, Eisenhower championed the Bureau of Public Roads' idea of replacing America's patchwork of byways with a modern system that, he felt, would chiefly support economic development, improve highway safety, relieve congestion, and reduce accident-related lawsuits. Eisenhower said the plan also could "meet the demands of catastrophe or defense, should an atomic war come"—a gripping fear in 1956. The result: a $50 billion federal project of 47,000 miles of new highways, 55,500 bridges, 104 tunnels, and 14,750 interchanges—and no traffic lights. These days, city stretches of the interstates can be congested nightmares, but overall the plan works. In 2005 more than 2 million Gulf Coast residents used interstates to flee hurricanes Katrina and Rita. And immediately after September 11, when planes everywhere were grounded, commerce continued, thanks to the interstate system.

50. Freedom of Speech:

Just don't yell "Fire!" in a crowded theater. Free speech, our most cherished freedom guaranteed by the First Amendment, gives us the right to say anything to anyone, anywhere, anytime—*except* if we are in a school, or if it is obscene, or if it is slanderous, or if it is libelous, or if it incites people to fight, or if it is harmful to minors, or if it is protected by a gag order, or if it is broadcast, or if it is hateful, or . . . You get the picture? This freedom has limits, but the U.S. is still better than anywhere else when it comes to having our say.

49. Apples:

Mostly Delicious. Washington State produces more than half the 10 billion pounds of apples grown in the United States each year, and Red Delicious is by far the most popular of the 2,500 varieties. Few family activities are more wholesome than apple picking in the autumn when the fruit ripens and people eat 'em from the tree. For the best apple pie, we suggest half Rome Beauty and half Granny Smith for tartness. Granny Smith originated in Australia, but is now one of the most popular varieties in America.

48. Local Government: *The people you call to fill in the potholes.* The politicians in Washington make the big decisions about the future of America, but try sounding off to *them*. In your own town or city, though, you can make a stink about bumpy streets, deteriorating schools, or dogs running loose, and somebody will actually listen to your complaint—and might even do something about it.

47. Broadway: *The Great White Way.* Millions of marquee lights along Broadway in Manhattan created the century-old nickname for the most famous theater district in the world. A must-see for tourists, Broadway shows, especially the musicals, have ratcheted up the spectacle—chandeliers fly across the stage (*Phantom of the Opera*), and teacups and candlesticks come alive (*Beauty and the Beast*). The astronomical cost of producing on Broadway has spawned a reliance on proven winners and safe choices, including retooled Disney animated features, and revivals, like the ever reliable Rodgers and Hammerstein's *South Pacific* or a limited run of the 1890 drama *Hedda Gabler* or the 1945 comedy *Blithe Spirit*. Only four theaters are actually located on the street Broadway, but Broadway counts 39 nearby theaters as part of its $1 billion annual industry. Theater snobs forego the mainstream dreck and seek the out-of-the-way and experimental offerings of Off Broadway and Off Off Broadway.

46. The Rockies: *Natural highs.* The mother of North American mountain systems, the Rockies cover some 300,000 square miles from northwest Alaska to Mexico, forming the Western continental

divide. Over 1,000 peaks soar above 10,000 feet. This gorgeous region boasts four national parks: Yellowstone, Grand Teton, Glacier, and Rocky Mountain. Some of our major rivers, including the Missouri, Rio Grande, Columbia, and Colorado, originate here where watersheds feed about a quarter of the U.S. water supply. Coronado, Lewis and Clark, Zebulon Pike, Sir Alexander Mackenzie, and Simon Fraser explored this terrain. But gold discoveries during the 1850s and 1860s led to permanent settlements and formation of the mountain states. The inspiration for the song, "Home on the Range," the Rockies still see "the deer and the antelope play." Now, however, the wildlife is joined by multitudes of campers, hikers, mountain climbers, and skiers.

45. Believing You Can Do Anything You Set Your Heart to:

Dream big! We hear stories like this all the time: Three impoverished kids from the inner city make a pact to help each other rise above their circumstances and become doctors. After they succeed, they compound their success by writing a best-selling book about their journey. American kids are taught to believe in the power of self—that dreams are attainable when we work really, really hard. Just look at the inspirational stories behind President Barack Obama, mogul Warren Buffett, comic actor Jim Carrey, astronaut Sally Ride, and U.S. Justice Antonin Scalia of the Supreme Court. Believing in themselves made all the difference.

44. The Bald Eagle: *Winged victory.*

In 1782, Congress ended six years of contention over the choice of a national emblem by rejecting Benjamin Franklin's nominee, the wild turkey, and choosing *Haliaeetus leucocephalus*. A big country needs a big bird. Let Nicaragua have its Turquoise-browed Motmot, all 2.3  ounces of it. The bald eagle is a soaring, imperious-looking bird of prey that can weigh up to 15 pounds and stretch its wingspan to almost eight feet. Endangered species status from 1967 to 2007 saved the eagle from extinction. The Indians believed that killing one brought bad luck. Now it's just illegal.

43. Conquest of the Moon: *Our gift to the world.* Yes, we planted the Stars and Stripes that summer day in 1969, yet we did not claim the moon for ourselves. When astronaut Neil Armstrong put his left foot down on the lunar surface, he said, "That's one small step for man, one giant leap for mankind." The astronauts also left behind a plaque that said they had come "in peace for all mankind." The landing fulfilled an ancient human fantasy, which found expression in the 19th century Jules Verne novel *From the Earth to the Moon,* and inspired grandiloquence as soaring as the poem that the *New York Times* published by Archibald MacLeish: "From the first of time / before the first of time, before the / first men tasted time, we thought of you. . . . Now / our hands have touched you in the depth of night."

42. Number 42: *Remembering Jackie Robinson's legacy.* To honor the first African-American Major League baseball player's courageous trailblazing, all the teams retired his number, 42, in 1997, the 50th anniversary of his achievement. Players of all colors in youth and school leagues still wear the number. They want to emulate a hero whose achievement transcends baseball.

41. T-shirts: *They're the tops.* Named for their simple design and initially worn as underwear, these white, cotton, short-sleeve shirts came out of the closet in the 1950s when actors John Wayne, Marlon Brando, and James Dean shocked audiences by wearing them solo—not under dress shirts—on the wide screen. They soared in popularity during the late 1960s and 1970s after rock bands and profes-

sional sports discovered the big money in having their names printed on T-shirts and advertising themselves on fans' torsos. Nowadays these tops—also available in polyester and every color and price under the sun—often voice the wearer's personal or political views. Still, pairing an inexpensive T-shirt with jeans has become the quintessential American uniform.

40. Communications Industry: *Changing with the* Times. The United States is the communications capital of the world: 124 million people read a newspaper every day; 220 million use the Internet; we each watch two months of nonstop television annually; 203 million of us carry cell phones, and the 69 million who text message send about 100 billion texts each month; 83 percent of us listen to the radio; more than 10,000 magazines are regularly published. We twitter and we blog. Communications is a multimedia industry in flux, but when the dust settles and newspapers are dust (it's just a matter of time), we predict that American media moguls will have figured out the next new thing to keep us in tune and in touch.

39. The Mississippi River: *It just keeps rollin' along.* The name comes from the Native American word Messipi, meaning big river. Though the Missouri is 240 miles longer, the 2,320-mile Mississippi is America's mightiest river, a highway of commerce and the country's mythic dividing line from Minnesota to the Gulf of Mexico, where it dumps its legendary mud. To the east, the safe, settled, and populous old America. To the west, the wide-open spaces where a man, it was said, could breathe free. It was a wild, sometimes raging monster when French explorer Sieur de La Salle sailed its length in 1682, and after decades of flood control projects, global warming threatens the river with floods like the monster in Iowa in 2008 that destroyed property worth billions.

38. California: *Golden opportunity for dreamers.* Born of the 1849 gold rush, California proclaims the weirdest state motto of them all: "Eureka!" (I have found it.) Found what? Gold? Natural beauty (the Sierra Nevadas, the Mojave Desert, the Redwood forests)? Big-shot status (a $2 trillion economy and 40 million people, to say nothing of a budget deficit in the eleven digits)? Cities of mystical allure (laid-back San Fran, frenetic L.A.)? Astonishing contrasts (north/south, Mount Whitney/Death Valley, Central Valley's agricultural bounty/Silicon Valley's technology)? Only-in-California status (home to half the country's Buddhists and to movie-star and body-builder governors)? Surely not its famous natural disasters (earthquakes, tsunamis, wildfires)? Wine? OK, that clinches it. Let's hit the road to Napa Valley.

37. American Pizza: *A slice above.* Our love affair with this street food started after World War II when GIs brought home a craving for the pies they had eaten in Naples, Italy. Today, pizza is the favorite national nosh. More than 3 billion pies are sold here annually; that's 23 pounds for each man, woman, and child. Total annual restaurant sales exceed $38 billion. Regional preferences are all over the map. Northerners favor the Neapolitan-style with its thin crust, tomato sauce, mozzarella, and simple toppings—pepperoni, the favorite. Chicagoans dig into deep dish pies; Californians toss their double-crusted numbers with anything from clams or pineapple to pesto and eggplant. And pies down South are laced with salsa, green chilies, jalapeños, and chorizo.

36. Television: *Exporting Barney, Jack Bauer, and Dr. McDreamy.* Television programming spreads the American way of life to even the most remote areas of the world. Characters like 24's Bauer, and Dr. Derek Shepherd, a.k.a. McDreamy, from *Grey's Anatomy*, provide riveting if unrealistic portrayals of hero-types in everything from government war rooms to hospital operating theaters. Children's shows like *Sesame Street* and *Barney* whet an appetite for a lifetime of television viewing. Americans watch TV an average four hours 35 minutes every day. While most

Baby Boomers grew up with a maximum of seven broadcast channels, the advent of cable television in the late 1970s triggered an explosion of specialized channels and programming. Baseball fans can watch the MLB Network, chefs the Food Network, crime story buffs *Law & Order* marathons on TNT, and music lovers MTV and VH1. American homes now have more televisions than residents and have access to an average 118 stations. This love affair with television extends far beyond our borders. American slang, habits, styles, trends, sports, and music have spread globally because of the influence of our television programming. Think *Desperate Housewives* or *Lost* is a little weird in English? Try watching in French!

35. Our Military Might: *The civilized world's 911.* For all the bashing America gets for throwing its weight around, our military's main purpose has been to maintain international peace and order. Sure, our motives have not always been pure (think of the 1846-48 war with Mexico and the 1898 Spanish-American War). But, mostly, we raise the sword not to colonize a country but to subdue wicked regimes like Imperial Japan, Nazi Germany, and Slobodan Milosevic's Yugoslavia. After helping rebuild them, we step aside. Europe often bridles at our supposed cowboy recklessness, but when danger looms, it gets on the phone.

34. The Bill of Rights: *People versus big government.* Drafters of the U.S. Constitution were determined to limit powers of the central government and wasted no wordage on rights of the people. However, they also designed this great document to be  subject to change as time passed. That foresight eased the way for addition of 10 amendments we commonly know as the Bill of Rights, the result of the 1787-88 debate on ratification of the Constitution. In the debate, the states argued successfully that the Constitution ought to include language providing civil liberties such as freedom of speech, press, religion, and the rights to bear arms and trial by jury. Today we may have 27 amendments, but the first ones, ratified in 1791, are as cherished and controversial as ever, judging from the cases heard by the U.S. Supreme Court every year.

33. National Park System: 84 *million acres preserved.* Since the creation of Yellowstone Park in 1872, the federal government has purchased and protected almost 400 sites, saving our natural heritage and giving future generations a sense of what pioneers saw. "Everybody needs beauty as well as bread. Places to play in and pray in, where nature may heal and cheer and give strength to body and soul alike," said 19th century naturalist John Muir, an early champion of preservation. Today, the largest of the government's holdings is Wrangell-St. Elias National Park and Preserve, an Alaskan tract almost as big as West Virginia. The smallest is the 900-square-foot Philadelphia house where Revolutionary general Thaddeus Kosciuszko lived. English-born Mary Bomar, the first naturalized citizen to head the National Park Service, recommended a visit to a national park as an antidote to American youngsters' lamentable addiction to video games. "Help them discover their America—without a joystick in their hands," she told parents.

32. Tough Building Codes: *Guardians against ghastly deaths.* Ours are probably the strictest in the world. Quake-threatened California has instituted such stringent measures that a powerful earthquake that roiled the Los Angeles area in 2008 did virtually no damage. After Hurricane Andrew devastated Florida in 1992, legislators scrambled to tighten laws. Stringent standards protect us from electrical and heating malfunctions, fires (think smoke alarms and multiple exits from buildings), and poison (requirements for testing and removing carbon monoxide, radon, asbestos, and lead paint). The list is endless; just ask your contractor. Penalties for violations can be severe, though not as harsh as the code of Hammurabi in ancient Babylonia: "If a builder has built a house for a man . . . and the house he built has fallen, and caused the death of its owner, that builder shall be put to death."

31. Worker Productivity: *Jobs "R" Us.* Our 153 million-strong workforce is the globe's third largest, after China and India. But American workers are still more productive per person than any others in the world, according to the International Labour Organization. In 2006 each U.S. worker produced $63,885 of wealth, well ahead of second-place Ireland at $55,986. And American employees proved themselves yet again despite the recession, increasing productivity in the third quarter of 2009 at an annual rate of 8.1 percent. No doubt about it, we owe our high standard of living to our nation's workers.

30. Religious Freedom: *A level praying field.* The Pilgrims and the Puritans fled England for the right to worship as they chose, though they punished dissent (as four Quakers hanged on Boston Common in the mid-1600s discovered). But as immigrants of all creeds poured into the Colonies, a tolerant mood took hold, finding its ultimate expression in the First Amendment's opening: "Congress shall make no law respecting an establishment of religion, or prohibiting the free exercise thereof." That didn't stop a mob from killing Joseph Smith, founder of the Mormons, but today, people of all faiths, from Catholics to Presbyterians, Jews to Jehovah's Witnesses, Evangelicals to Episcopalians, Muslims to Mennonites, and hundreds of other sects, denominations, and movements, are free to praise God as they please—or skip services along with agnostics and atheists, who are equally free to deride belief in the Almighty.

29. Scientific Achievements: *Envy of the world.* We are the uncontested leader in research in just about every scientific discipline you can think of. As of this writing, Americans have won close to 60 percent of Novel prizes awarded in medicine, physics, and chemistry (capturing all three in 2009). We dominate Internet, energy, and space technology, nanotechnology, engineering, genetics, and transistors. Our universities and other research centers attract

hordes of foreigners to our shores. Our supremacy probably won't last forever, but that's OK: the human race needs all the help it can get to survive into the 22nd century.

28. Hawaii: *Land of superlatives.* It is the most gorgeous, most exotic, most southern, most isolated, and widest state, spread out over 1,500 miles. It was the last admitted to the Union. It has the largest percentage of Asian Americans and mixed-race people. It is the only state not located in North America, the only that was once a kingdom, the only surrounded by water, the only with two official languages, the only that grows coffee, the only with its own time zone. It is home to the world's biggest telescope.

27. Thanksgiving: *America's holiday.* Horse-drawn sleighs used to travel over the river and through the woods to grandmother's house, but now 38 million Americans take to the roads, air, and rails to observe this annual day of thanks. Thanksgiving began in 1621 when Pilgrims and Native Americans feasted for three days after the harvest. There's no hard evidence that turkey was the center of the festivities then, but historians muse that, given the proximity to the ocean, lobsters were probably plentiful on the tables. President Lincoln declared the day a national holiday. Football came later—in the 1890s. Macy's Thanksgiving Day Parade with its humongous helium balloons began in 1924. Now, some 47 million people watch it live or on television. Let's also pause and pay homage to the 46 million turkeys who annually give their lives for their country. Pass the cranberries, please.

26. Breadbasket to the World: *Food, glorious food.* The United States is a leading supplier of food to other countries because American farmers are the most efficient producers. Blessed with climate and soil conditions that foster cultivation of foodstuffs like

wheat, soybeans, corn, pork, and dairy products, they also have the advantage of advanced transportation, marketing, and financial infrastructures. Top markets are Canada, Mexico, Asia (except for Japan), and the European Union. Our farm exports tallied a record $115.5 billion in fiscal 2008, a bright spot in a trade picture that shows huge and endless deficits. And, with just six percent of the world's population, U.S. farmers, some years, provide 60 percent of foreign food aid.

25. "America the Beautiful": Song that should be the national anthem. Even professionals have trouble singing the "The Star-Spangled Banner," set to the tune of an old drinking ditty. Detractors of poet Francis Scott Key's tribute to American victory in the War of 1812 cite its warlike nature. Better, they say, to celebrate the spacious skies, amber waves of grain, purple mountain majesties, fruited plains. The words to "America the Beautiful" were penned in 1893 by a professor of English as she traveled by train from sea to shining sea.

24. Old Glory: Symbol of patriotism of all stripes. Our Constitution forbids monarchy, but we venerate colors more royal than purple: the red, white, and blue of our flag. It flies from millions of poles, a ubiquity that perplexes many foreign visitors. The 50 stars represent the 50 states, and the 13 stripes the original 13 colonies; the colors have no official meaning but many Americans take red to stand for valor, white for purity and freedom, and blue for vigilance and justice. We pledge allegiance to the flag as a symbol of freedom and national unity, and we sing "The Star Spangled Banner," our national anthem, which celebrates the banner's survival of a British bombardment of an American fort in 1814, at football games. Old Glory flew at the battles of Valley Forge, Gettysburg, San Juan Hill, the beaches of Normandy, Okinawa, and Khe San. As the Johnny Cash song says, "She's been through the fire before, / And I believe she can take a whole lot more." The flag flies at half-staff to honor our illustrious dead. It flaps next to the graves of soldiers on Memorial Day. It stands on the moon.

23. Cybernation: *Epic epoch.* It happened so fast. We became a population of cyber nuts in the blink of an iPod, a cultural shift that makes all those other social upheavals of the 20th century—automobiles, TV, space travel—seem like infatuations that preceded the love of our life. This romance is the real thing because it keeps evolving. We made a list of Internet-age breakthroughs ("blogging," "Facebook," "Google," "text messaging," "Twitter," to name a few), but stopped at 100 because by the time you read this, they will all seem so yesterday.

22. The Metropolitan Museum of Art: *Big is beautiful.* This 2-million-square-foot Gothic Revival style building in Manhattan's Central Park houses one of the most exhaustive collections anywhere—more than 2 million works of art spanning 5,000 years of world culture. There's something for everyone: Greek and Roman statuary, medieval tapestries, armor, Asian art, and European paintings including Rembrandts, Vermeers, and Impressionists. There's even an entire Egyptian temple, circa 15 B.C. The American Wing alone offers the world's most comprehensive collection of American paintings, sculpture, and decorative arts. More than 5 million visitors flock here each year, and most get lost in its cavernous galleries. On weekend nights the marble mezzanine becomes an elegant bistro. You haven't lived until you've sipped wine there, surrounded by lilting chamber music and some of the finest art in the world.

21. Jazz: *Soul of American culture.*

The spontaneity, improvisation, and resiliency that mark this musical form fuse the defining elements of the American character. Most jazz relies on the principle that an infinite number of melodies can fit the chord progression of any  song. When trumpeter Louis Armstrong, the first true virtuoso jazz soloist, was pressed to explain what this sound was, he responded: "If you gotta ask, you'll never know." Jazz was born in New Orleans in the early 1900s when black musicians combined principles and elements of European and African music. The style spread to subcultures in New York, Kansas City, Chicago, and St. Louis. Band leaders Duke Ellington and Fletcher Henderson embraced it, and musicians Miles Davis, Charlie Parker, Dizzy Gillespie, and Thelonius Monk played their hearts out. By the 1960s, jazz was considered an important musical form. Best place to catch it? Back home in New Orleans.

20. Entrepreneurship: *The secret to America's economic miracle.*

As much as mom and apple pie, it is the little guy with the big idea that defines us. Our grassroots capitalist culture is a wellspring of innovation, a lure to immigration, and our most effective weapon in the competition for world markets. No other nation comes close to matching our army of 25 million entrepreneurs, whose start-ups and fast-growing businesses account for half of our gross national product and two thirds of new jobs. Shunning authority, seizing opportunities nobody else notices, taking make-or-break risks and shrugging off failure, entrepreneurs embody the frontier spirit that transformed an upstart band of British colonists into the world's unquestioned superpower. More than politicians or movie stars, more even than sports heroes, it is entrepreneurs like Microsoft's Bill Gates, Apple's Steve Jobs, and Google's Sergey Brin who fire our imaginations with their creative destruction of old ways of doing things.

19. **Our Optimism**: *The glass is always half full.* It must have been a foreigner who quipped, "The light at the end of the tunnel could be another train." Americans like the sunny side. President Kennedy, for example, noted that seven percent unemployment meant 93 percent had jobs! We cheerfully assume problems are surmountable and that the good guy always wins. Who else but an optimistic American would have the audacity to take out a 30-year mortgage at age 72? We clearly win the international optimism contest—after all, we're the country that gave the world the ever-optimistic Pollyanna, Forrest Gump, and Disney World, the happiest place on earth.

18. **Our Generous Spirit**: *Digging deep for the deserving.* We are the most generous people on the planet. We rush food, blankets, and medicines to overseas victims of catastrophes or wars. We open our wallets to help domestic towns recover from hurricanes, floods, or tornadoes. We spearhead fund-raisers for the kid who needs a bone marrow transplant. In fact, charitable donations in the United States annually amount to more than one percent of the country's gross domestic product, reports Giving USA Foundation. That's better than the United Kingdom, the international second with 0.73 percent of its GDP earmarked for charities, and far ahead of France, Germany, Ireland, and Turkey. We've got more nonprofits than anybody else: 1.4 million, ranging from hospitals and human service organizations to advocacy groups and chambers of commerce. In prosperous times the average American household donates about two percent of adjusted gross income to charity. Surprisingly, families earning $100,000 or less a year are proportionately more generous than the rich. Of course, recession always puts the brakes on how much can be doled out: estimated total giving in 2008 was $307.65 billion, down two percent from 2007. Still, it's heartening to note that charitable donations don't decrease as badly as economic downturns would seem to dictate. "People still care, so do foundations and corporations," says Giving USA's Web Site.

17. Blue Jeans: *Classless denim swath that cuts across all demo-graphics.* Clothing manufacturer Levi Strauss used a nearly inde-structible blue denim fabric and secured his seams with copper rivets to fashion one-dollar work pants for California miners during the 1849 Gold Rush. Somewhere along the way they turned into a fashion statement that can cost thousands of dollars. You still can get a pair of plain old Levi's for about $40.

16. The Great American Novel: *Chronicle of change.* Want to understand U.S. his-tory without studying boring dates? Read the canon of prose fiction and see through the eyes of characters trying to cope with the changes each generation has endured. It started with James Fennimore Cooper's *Deer-slayer* novels of the new frontier, and was fol-  lowed by Harriett Beecher Stowe's *Uncle Tom's Cabin,* with its cry against slavery, and Mark Twain's whimsical tales of the 19th cen-tury heartland. In *Sister Carrie,* Theodore Dreiser wrote of the rise of the cities. F. Scott Fitzgerald and Thomas Wolfe told of the longings of the Jazz Age, John Steinbeck's *The Grapes of Wrath* became a primer on the Great Depression, and Jack Kerouac's *On the Road* sang of the emptiness of the post–World War II spirit. Among the great contemporary novels, Don DeLillo's *Underworld* led readers through a search for meaning—if there was any—in a mechanistic modernity. Not a bad syllabus for American Lit 101.

 **15. Hamburgers:** *The king of cuisine.* How many hamburgers do Americans eat each year? Let's say a zillion. You got a better number? Some food historians say a burger on a bun was invented at the 1904 World's Fair, but no one can prove it or say for sure why it is named after a German city. Alton Brown, host of the *Good Eats* show on the Food Network, recommends half sirloin, half chuck, for what he calls the "Burger of the Gods." It's a mistake to order one

abroad. Try it in a tourist restaurant in Madrid and you'll get a piece of ham. Yuck. Nutritionists hate chopped beef because too much fatty red meat is bad for the cardiovascular system. But a burger with cheese, tomato, and onion is one of the best things about America because it tastes so good.

14. Public Libraries: *Democracy's bookshelf.* Since the first one opened in 1656 in Boston, public libraries have grown to 16,000 branches dispensing 2.1 billion items a year. Nineteenth century steel baron Andrew Carnegie, who believed that libraries gave citizens the information they needed to excel, bestowed a huge fortune to create free public libraries around the world, 1,679 in the United States. Today, bibliomania appears to be strongest in Ohio, where the average resident visits a library 7.2 times a year. For the dwindling few who lament the passing of the white-haired librarians who hissed *Sssshh,* we say get with it, and learn to tap into the electronic information resources at most libraries. The coming big thing is access to computer gaming, which most librarians favor as a way of attracting younger patrons, according to the American Library Association.

13. The Grand Canyon: *Making humanity insignificant.* "Oh my God!" is the most common phrase among visitors getting their first glimpse from Mather Point at the southern approach in Arizona. The canyon began to emerge 5 or 6 million years ago as the Colo-

rado River burrowed into the earth and then receded, leaving a hole rimmed with rocks 277 miles long, a mile deep. "Ours has been the first, and will doubtless be the last, party of whites to visit this profitless locality," explorer Joseph Ives said in 1858. Now, the 5 million awed visitors who come each year threaten the ecology. The year 2016 is the Park Service's target date for finding "green" replacements for those polluting gas engines on rafts in the river.

12. The Melting Pot: *Assimilation's metaphor.* It's been bubbling ever since straitlaced Pilgrims sat down to dinner with the Native Americans in Plymouth. More than 65 million have come: English, Germans, Italians, Irish, Russians, Slavs, Cubans, Vietnamese, Mexicans, Japanese, and Chinese. The Africans came under duress. Others came to escape tyranny, religious persecution, or to create a better life. They built our infrastructure—railroads, skyscrapers, canals, tunnels, and roads. Their children learned English in school. Embarrassed by Old World parents, second generationers do everything they can to look and feel like Americans. Voilà! Assimilation! Martin Luther King, Jr., summarized the phenomenon: "We may have all come on different ships, but we're in the same boat now."

11. Class Mobility: *The economy's escalators.*
Cornelius Vanderbilt, a farmer's son who left school at 11, became the richest man in 19th century America. Oprah Winfrey, reared on a farm by her grandmother and sent to a juvenile detention center at 13, blossomed into the nation's most popular TV talk show host in the 20th and 21st. The ride down can be as fast as the ride up, as the victims of Bernie Madoff—the disgraced pyramid scheme scam artist—discovered in 2009.

10. Wide-Open Spaces: *Our ticket to the American Dream.* The Old World was constricted, parceled out, every acre hoed and fought over for 2,000 years. The New World was unexplored, a vastness of forest and prairie, mountains, lakes, and rivers. Thomas Wolfe, an early 20th century American novelist known for his exuberant

prose in such masterpieces as *Look Homeward Angel,* woke up one morning in the "wool-soft air of Europe," aching for "the wilderness, the howling of great winds, the bite and sparkle of the clear, cold air" of America. We feel that same pang; like Huck Finn, we can't wait to light out for the territory—which the average American, seeking a new start, does 11.7 times in his or her lifetime. Why not? The vastness is still out there, beyond the malls and the superhighways, all 2.4 billion acres of it, about 7.8 acres for every inhabitant, man, woman, and child.

9. Baseball: *The perfect game.*

What compels 78 million people a year to pay admission to watch millionaires swat cowhide spheres with sticks? Why does a mother praise God when her five-year-old finally makes a catch? And why does a son place a tattered mitt, signed by Detroit Tiger Mickey Cochrane in 1930, beside the Rosary and Purple Heart in a father's coffin? The answer: love of a pastime whose memories bind America's generations and whose green fields remind us of our rural heritage. The rules, mostly intact after 150 years, require a balance of ballet, athleticism, and brainpower. The sport teaches the ingredients for group success: individual achievement within a team. And baseball is like life itself, as the late baseball commissioner Bart Giamatti wrote: "The game begins in the spring, when everything else begins again, and it blossoms in the summer, filling the afternoons and evenings, and then as soon as the chill rains come, it stops and leaves you to face the fall alone."

8. Rock 'n' Roll: *It's here to stay!*

Stodgy, sing-along band leader Mitch Miller scoffed that rock 'n' roll wasn't music, it was "a disease." Ha! Nearly 60 years after Cleveland DJ Allen Freed described this new genre, rock 'n' roll has shaken, rattled, and rolled over other formats. Grunge, punk, new wave, pop, hard rock, heavy metal, and emo live harmoniously under its broad umbrella. Rock, with a 32.4

percent share of the U.S. music industry, has given the world Elvis Presley, the Beach Boys, Chuck Berry, the Doors, the Grateful Dead, Grand Funk Railroad, Chicago, Green Day, the Red Hot Chili Peppers, and the Foo Fighters. If that's disease, it's highly contagious. As Billy Joel sang, "Hot funk. Cool Punk. Even if it's old junk, it's still rock 'n' roll to me."

7. New York City: *From $24 bargain to world's greatest metropolis.*

What the Dutch bought from the Indians for a few trinkets in 1624 is now the business and cultural capital of America. It tells the country how to think, what to buy, and what to wear. It has the best skyline, the best restaurants, and its 8.2 million people include immigrants from every country. The Jackson Heights neighborhood in Queens is home to 70 nationalities. Life can be hard. Brooklyn-born novelist Henry Miller said, "Every bloody street I look down I see nothing but misery." But the Big Apple is the place to be if you have brains and grit. As the singer says, "If I can make it here, I can make it anywhere." The energy and creativity are unmatched. Blind and deaf Helen Keller wrote of visits to New York, "Always I return home weary but I have the comforting certainty that mankind is real and I myself am not a dream." Comedian Jackie Gleason joked, "When you leave New York, you're camping out."

6. Religious Faith: *Glory, Glory Hallelujah!*

You can tell a lot about a nation by what is printed on its currency. The old French francs trumpeted "Liberté, Egalité, Fraternité" (and under the collaborationist Vichy regime "Work, Family, Country"), but also warned counterfeiting was punishable by life in prison. German marks were copyrighted. American bills declare, "In God We Trust." The phrase defines us. The Pilgrims left for the New World in 1620 so they could pray in peace. Our founding document, the Declaration of Independence, invokes God four times—as "Nature's God,"

"Creator," "Supreme Judge," and "divine Providence"—to justify America's break with Britain. Abraham Lincoln, America's Christ figure, proclaimed Thanksgiving as a day of "praise to our beneficent Father." Our Pledge of Allegiance reminds us that we are "one nation under God." Believers crowd our churches and synagogues. Religious freedom makes our worship possible. Faith makes it real.

5. The Constitution: *Democracy's mansion.* The 1789 blueprint for American government had a novel design, and chief architect James Madison called it "a system which we wish to last for ages." But there were no guarantees the Constitution would serve a growing country. So far it has. Two centuries of expansions, repairs, and upgrades have turned the starter home of U.S. democracy into a palace that is the model for the free world. The glittering facade is the Bill of Rights and the 14th Amendment's guarantee of due process of law. The plumbing and wiring are the duties and rights of the states and the branches of government. There are musty rooms too, that even grown-ups fear to enter, like the Electoral College, where the scurrying of mice is heard every four years. Our leaders swear an oath to preserve, protect, and defend this edifice. And when the owners disagree about maintenance, they call in the Supreme Court to paste sticky-note instructions on the walls.

4. The Movie Industry: *That's entertainment!* The U.S. film industry is the oldest and largest in the world. Italy has its spaghetti westerns, India has its Bollywood productions, which are making international waves, but the epicenter of the global film industry is Hollywood, California. That's where most of the movie blockbusters are born, in studios like Universal and MGM or through independent film makers like Spike Lee, Quentin Tarantino, or other lesser known directors. The industry generates about $35 billion in annual revenue, mostly through theater attendance

and DVD sales. Movies like *Avatar, Titanic*, the *Lord of the Rings* series, the *Star Wars* saga, and the *Batman* epics have generated more than a billion dollars in box office revenue. America's superstars—like George Clooney, Angelina Jolie, Brad Pitt, and Julia Roberts—are instantly recognizable the world over. The great movies have informed our popular culture, expanded our lexicon, and painted an interesting (but clearly inaccurate) picture of life in America for the international audience. We have never witnessed a car chase, don't know any bad guys, have never stayed in a haunted house, limit ourselves to one glass of wine each evening, avoid space and time travel whenever possible, and hardly ever say anything worth quoting ad nauseam. Really.

3. Visual Bounty: *This land is your land.*

There are the coasts, from the rock-strewn waters of Maine to the sun-drenched sands of southern California, 12,500 miles of shoreline in all. Then there is everything in between: the redwood forests, the Grand Canyon, the Rockies, the Mississippi, Niagara Falls, the national parks, the spacious skies, the fields of grain. British diplomat Sir Roy Denman, a champion of European integration and Europe's ambassador to Washington in the 1980s, marveled at the grandeur of the Utah Salt Flats, urging foreign visitors to drive through them for a taste of the vastness of the United States. French tourist Evelyne Laffargue, an artist and teacher who made her maiden voyage to the U.S. in 2004, gasped and clapped her hands the first time she glimpsed the Manhattan skyline as she boarded a ferry in Hoboken, New Jersey, that would transport her across the Hudson. Naming all the splendors would fill this book; visiting just the most spectacular of them would take a lifetime.

2. We the People: *Freedom's guardians.*

We—you, me, and all the other Americans—are the sovereigns of our country. The succinct Preamble to the Constitution begins with, "We the People" to underscore who rules—not the politicians or judges in Washington, D.C., not the states, not city mayors or town planners. When threatened by government excesses or abuses, we can always vote the bums out. We are the ultimate source of power, we alone have the right to alter the Constitution and even overthrow the government. Framer James Monroe lauded us as "the people, the highest authority known to our system, from whom all our institutions spring and whom they depend . . ." Truth be told, the Founding Fathers were wary of our influence if left unfettered. "In questions of power then," wrote future President Thomas Jefferson, "let no more be heard of confidence in man, but bind him down from mischief by the chains of the Constitution." Yet, if there is one thing that sets us apart from other democracies to this day, it is governance by institutions that are checked by popular forces. The people's supremacy is the principle that has bound this union together from the creation of the United States.

1. Pursuing the American Dream:
Yours, mine, and ours. First there was the land, rich with resources and natural beauty. Then came the people, and whether they were here for millennia or crossed the seas in the *Mayflower* in 1620 or flew in a jet from China a few weeks ago, they were the beneficiaries of a golden idea: the people could choose their own leaders and live in a society with a balance of freedom and order. Their history chronicled a struggle to give each individual the opportunity to make personal dreams come true. To be rich, well fed, and educated. To win an Oscar, fly to the moon on a rocket, or sit in a canoe and drink a beer. Not that America is paradise. You can't always get what you want. And we'll skip the details of the unfulfilled promises and the tarnishes on the dreams of the nation at large. The to-do lists are long and as varied as the country's races, regions, classes, cultures, incomes, and political leanings. In the meantime, the plus side of the ledger looks good. The 250,000 people who apply to be citizens each year (and the millions who sneak in) must think so too. The best thing about America is the

spectacle of its 310 million people trying to live honest and abundant lives and striving to make their individual and collective dreams real. To give meaning to the creed that all people are created equal and are endowed by their creator with the right to life, liberty, and the pursuit of happiness. To raise holy hell against injustice, as folk singer and writer Woody Guthrie urged. To keep the land as sweet as philosopher Henry David Thoreau found it at pristine Walden Pond in Massachusetts in 1854. And to act on President Ronald Reagan's admonition: "Freedom is never more than one generation away from extinction. We don't pass it to our children in the bloodstream. It must be fought for, protected, and handed on for them to do the same."

Your Chance for
Fifteen Minutes

of Fame

Okay, so you think we goofed.

Tell us what we foolishly left out, and why. We may put your suggestion on our Web site or in a future book. If we do, we'll acknowledge your contribution.

E-mail us at Authors@1000ThingsToLoveAboutAmerica.com, and, in approximately 150 words, tell us why the thing you love about America is important.

Please visit us at www.1000ThingsToLoveAboutAmerica.com.

ACKNOWLEDGMENTS

WE WOULD LIKE TO THANK the many, many family members, friends, acquaintances, colleagues, and sundry strangers we met at parties and in bars who graciously and sometimes aggressively gave us their opinions about what we ought to include in this book. A partial list follows. So many people weighed in that we have almost certainly omitted the names of quite a few people who, we confidently predict, will let us know of our lapse. On the other hand, a small number of family members, friends, and acquaintances ignored our appeals for input and we are pleased to withhold their names.

Here goes for the folks who answered the call: most especially, our children, Jennie Bowers, Matthew Zowers-Bowers, Annie Gottlieb, Daniel Gottlieb and Thomas Hooper Gottlieb; also, Barry Adler, Kathleen Bowers, Olive Bowers, Robert E. Bowers, Fred Brock, Frank Csongos, David Dorn, Abby Ellin, Christopher Elliott, Bela Gajary, Perry Garfinkel, Kurt and Monica Hantusch, James Haverkamp, Susan Hoogsteden, Chuck Jackson, Peter Johnson, Margaret Loke, Joan O'Neill, Ronald Panko, Francine Parnes, Pat Permakoff, Larry Petraccaro, Harry Rijnen, Terry Rooney, Theodore Stanger, David Tuck, Stephen Wasnok, Ana Westley, Dr. Thomas Whelan, Robert Wielaard, Tessa Wilkinson, Larry and Donna Young, Bernadette Hooper, Peg and Bob Huryk, Jim Huryk, Maryrose Huryk, Robert Huryk, Maryrose and Gene Mangan, Alyson and Andrew Flanagan, Maureen and Mike Madden, Jennie and Jon Harris, Therese and Don Thompson, Tom and Darla Hooper, John and Karen Hooper, Martin and Peg Hooper, Jim and Marilyn Hooper, Matt Hooper, Bernadette and John McVey, John McVey Jr., Tom McVey, Emily McVey, Deirdre and Dan Yates, Moira Donohue and

Rob Tobiassen, Dr. Andrew Weinberger, Bernadette Manno, Forrest Pritchett, Elizabeth Hoehn, Robin Cunningham, Joan Brennan, Majid Whitney, Hezal Patel, Elizabeth Cappelluti Sheehy, Matt Geibel, Amanda DiDonato, Chris Breen, Elizabeth Stevens, Carla Stevens, Catherine Stevens and Jeff Gluckman, Pat and Stephen Gluckman, Fern Denney, Jay Fuhrman, Carly Fuhrman, Richard Gottlieb, Kate and Paula Hurley, Kitta MacPherson, Megan Hall, Tom Duffy, Maureen Miller, Pamela Brownstein, Robert Stack, John Erway, Gianna Kenny, Candida DaFonseca, Lynn and Tom Benediktsson; the Friday night dinner gang, and that man from the Philippines in the San Francisco airport who said Americans, of all the world's people, were the friendliest.

We also thank our agent, Robert Wilson, and our editor, Kate Hamill. Their guidance and suggestions made this project what it is.

ABOUT THE AUTHORS

Barbara Bowers, a former NBC Radio news writer, Gannett newspaper reporter, and editor at *Best's Review* magazine, now is a freelance writer. A former reporter and editor at the *New York Times* and the *Wall Street Journal*, Brent Bowers writes for the *Times* and is the author of several books. Agnes Hooper Gottlieb, Ph.D., a former *Associated Press* reporter, is the Dean of Freshman Studies at Seton Hall University. Henry Gottlieb, a former *Associated Press* reporter and foreign correspondent, is a writer for the *New Jersey Law Journal*. Together, these four journalists wrote *1,000 Years, 1,000 People: Ranking the Men and Women Who Shaped the Millennium*.